Soil Carbon: Science, Management and Policy for Multiple Benefits

Soil Carbon: Science, Management and Policy for Multiple Benefits

Editors

Bhaidas Patil, Ravindra Borade

scitus academics

Soil Carbon: Science, Management and Policy for Multiple Benefits

Edited by **Bhaidas Patil, Ravindra Borade**

ISBN: 978-1-68117-094-7

Library of Congress Control Number: 2015935493

SCITUS Academics LLC,
www.scitusacademics.com
Box No. 4766, 616 Corporate Way,
Suite 2, Valley Cottage,
NY 10989

Printed in United States of America on Acid Free Paper ∞

Contents

Preface

Managing Soil Organic Carbon for Global Benefits presents an overview of current technical and scientific knowledge of soil organic carbon (SOC). Such an overview is needed in order to understand how and why soil organic carbon management should be an important component of future strategy for the Global Environment Facility (GEF). SOC management may contribute to and/or be affected by progress toward the achievement of the GEF objectives. It has a vital role to play in regard to each of the GEF focal areas: Biodiversity, Climate Change, Land Degradation, International Waters, and Persistent Organic Pollutants (POPs). Links also exist between SOC management and the GEF's cross-cutting themes (Sustainable Forest Management, and Land Use, Land-Use Change and Forestry – LULUCF). Knowledge of SOC and SOC dynamics has grown considerably in recent years, driven by a number of current debates including on the potential of soils to contribute to climate change mitigation, ecological approaches to agriculture, and pro-poor agricultural development, among many others. SOC monitoring could benefit from remote sensing techniques developed in recent years. However, increased knowledge has confirmed that SOC dynamics are difficult to assess and predict – not only because SOC is complex in itself, but also because the inter-relationships between SOC and its controls are complicated. Better knowledge can therefore rarely be directly translated into straightforward SOC management rules; rather, this knowledge contributes to understanding a complex system whose response is inherently difficult to predict, yet where multiple benefits can potentially be achieved.

Editor

Chapter 1

The Potential for Carbon Sequestration in Australian Agricultural Soils is Technically and Economically Limited

Shu Kee Lam, Deli Chen, Arvin R. Mosier, and Richard Roush

Melbourne School of Land and Environment, the University of Melbourne, Victoria 3010, Australia.

ABSTRACT

Concerns about increasing concentrations of greenhouse gases in the atmosphere, primarily carbon dioxide (CO_2), have raised worldwide interest in the potential of agricultural soils to be carbon (C) sinks. In Australia, studies that have quantified the effects of improved management practices in croplands on soil C have generally been inconclusive and contradictory for different soil depths and durations of the management changes. We therefore quantitatively synthesized the results of Australian studies using meta-analytic techniques to assess the technical and economic feasibility of increasing the soil C stock by improved management practices. Our results indicate that the potential of these improved practices to store C is limited to the surface 0–10 cm of soil and diminishes with time. None of these widely adopted practices is currently financially attractive under Australia's new legislation known as the Carbon Farming Initiative.

Subject terms:

INTRODUCTION

Globally, soils are estimated to contain 1500 Pg of soil organic carbon (C) to a depth of 1 m, approximately double the amount of C in the atmosphere[1]. Around 160 Pg of organic C are stored in soils of agricultural croplands in the world[1]. A meta-analysis showed that long-term conversion of native forest and pasture to cropland has reduced soil organic C stocks by an average of 42% and 59%, respectively[2]. By adopting improved agricultural management practices, agricultural soils in the world are estimated to have the potential to sequester 0.4–0.8 Pg C per year[3].

Improved agricultural management practices such as conversion from conventional to no-till or reduced tillage, residue retention, conversion to permanent pasture, crop rotation and fertiliser application have been shown to increase soil C in various countries, including Australia[4, 5, 6]. However, contradictory results are common in the literature. For example, a meta-analysis based mostly on studies in North America and South America indicated that adoption of no-till did not increase soil C stock down to 40 cm (ref. 7), and the potential for C accrual following conversion from crop to pasture decreased with soil depth[2]. Carbon sequestration rates under different grassland management practices have been demonstrated to vary substantially over time[4].

Large and continuous inputs of organic matter are conducive to increasing soil C content[8]. The net primary productivity of most of Australia's agricultural land, which determines the potential C input, is water and nutrient limited[9]. Soil C concentrations in Australia are inherently low, except for limited areas under higher rainfall along the east coast[5]. The range of soil C in the upper 1 m of soil in Australia is 18–447 Mg ha^{-1}, compared to a global value of 30–800 Mg ha^{-1} (ref. 8). While the impacts of agricultural management practices on soil C sequestration potential in Australia have been reviewed[5, 9], a quantitative synthesis of the responses of the soil C stock to these practices for a range of soil depths and years of implementation is lacking. This information is important for assessing the potential of improved management practices to increase C storage, or reduce C loss, within soil profiles and over time. Farmers and land managers need this information to determine whether they can earn carbon credits under Australia's Carbon Farming Initiative (CFI), which was launched to create a financial incentive for the adoption of management practices that store soil C permanently (100 years or more) and improve the soil condition and hence its productivity[10].

The carbon and nitrogen (N) cycles are tightly coupled[11]. The C: N ratio of agricultural soils generally maintains a constant narrow range, with an average of 10:1 [8]. Thus, N is required to stabilise agricultural soil C accrual[6]. However, the cost of N for C storage is often overlooked in assessing the benefit of C credit earned from improved management practices. We therefore conducted a meta-analysis of published data on the responses of soil C to improved agricultural practices, viz. conservation tillage, residue retention, the use of pasture, and fertiliser N application. The objectives of this analysis were (i) to quantify the changes in soil C concentration and soil C stock under improved management practices, with respect to soil depth and management duration; and (ii) to provide a simple assessment of the economic feasibility of these practices.

RESULTS

Effect of Improved Management Practices on Soil C at Various Soil Depths

Soil C concentration in the surface soil layer (0–10 cm) was significantly greater under conservation tillage (by 10%), residue retention (by 8%), the use of pasture (by 13%) and fertiliser N application (by 7%), when compared to control treatments (Fig. 1). Soil C stock in surface soil (0–10 cm) was higher under these management treatments than in the untreated controls. The relative gain in soil C stock in this surface layer was in the order of: pasture use (140 kg C ha^{-1} $year^{-1}$) > conservation tillage (139 kg C ha^{-1} $year^{-1}$) > residue retention (62 kg C ha^{-1} $year^{-1}$) > fertiliser N application (47 kg C ha^{-1} $year^{-1}$) (Table 1). Neither soil C concentration nor soil C stock was significantly increased by these management in soil deeper than 10 cm (Fig. 1; Table 1).

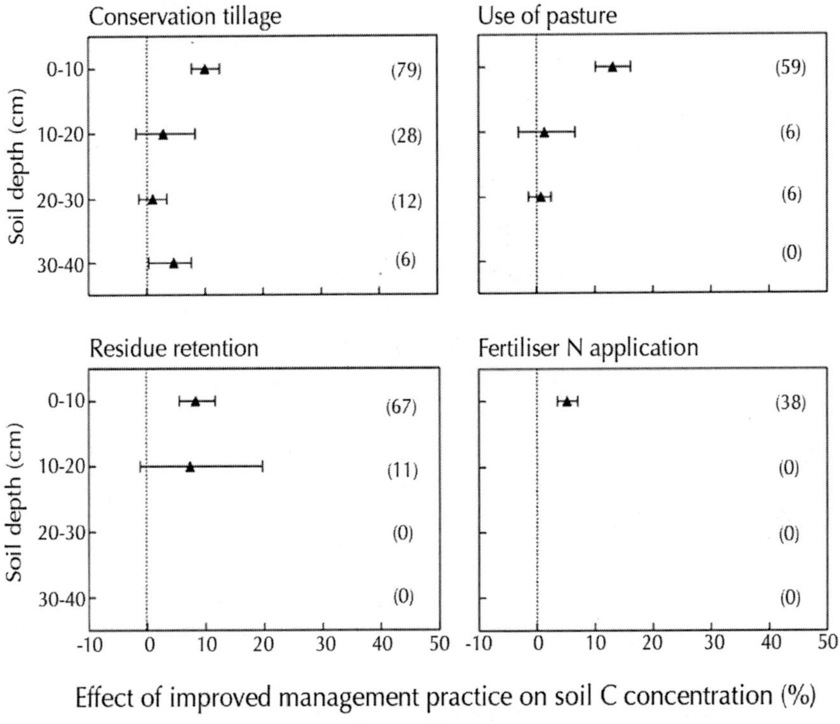

Figure 1: Effect of improved management practices on soil C concentration at various soil depths.

Means and 95% confidence intervals are depicted. Numbers of experimental observations are in parentheses. The database of the meta-analyses can be found as Supplementary Tables S1–S4 online.

Table 1: Effect of agricultural management practices on soil C stock at various soil depths. The database of the meta-analyses can be found as Supplementary Tables S1–S4online

Soil depth (cm)	Relative change in C stock (kg C ha^{-1} year^{-1})	
	Mean	95% CI
Conservation tillage		
0–10	139	101 to 184
10–20	24	−27 to 75
20–30	21	5 to 35
30–40	36	0 to 80
Residue retention		
0–10	62	31 to 91
10–20	30	−5 to 71
20–30	2	−9 to 11
30–40	NA	NA
Use of pasture		
0–10	140	69 to 250
10–20	−18	−153 to 128
20–30	14	−60 to 65
30–40	NA	NA
Fertiliser N application		
0–10	47	28 to 66
10–20	11	1 to 20
20–30	2	−8 to 11
30–40	NA	NA

Effect of Improved Management Practices on Soil C under Different Experimental Durations

The four management practices significantly increased soil C concentration for at least 20 years after the treatments began (Fig. 2). These increases were averaged across soil depths, but were reflective of the changes occurring mostly in the 0–10 cm layer (Fig. 1). While the effects on soil C concentration of residue retention (2–10%) and fertilizer N application (3–7%) decreased with time, those for conservation tillage (7–10%) and pasture use (10–13%) were sustained longer (Fig. 2). The annual relative gain in soil C stock was greatest during the first 10 years after implementation of improved management treatments, when the increases were in the order of: conservation tillage (150 kg C ha^{-1} $year^{-1}$) > residue retention (147 kg C ha^{-1} $year^{-1}$) > pasture use (132 kg C $ha^{-1}year^{-1}$) > fertilizer N application (67 kg C ha^{-1} $year^{-1}$) (Table 2). These increases correspond to a soil C gain of 247–552 kg CO_2-equivalent (CO_2-e) ha^{-1} $year^{-1}$ during the first 10 years of implementation of the improved management practices (Table 2).

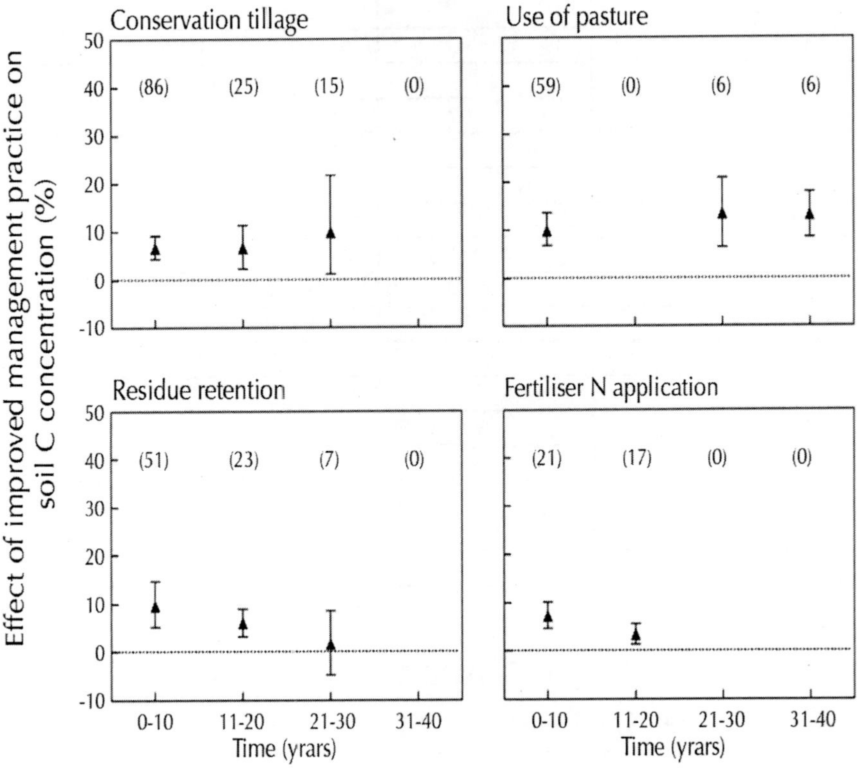

Figure 2: Effect of improved management practices on soil C concentration under different experimental durations.

Means and 95% confidence intervals are depicted. Numbers of experimental observations are in parentheses. The database of the meta-analyses can be found as Supplementary Tables S1–S4 online.

Table 2: Effect of agricultural management practices on soil C stock, C credit and financial returns under different experimental durations. The database of the meta-analyses can be found as Supplementary Tables S1–S4 online

Duration (year)	Relative change in C stock (kg C $ha^{-1}year^{-1}$)		CO_2-equivalent (kg CO_2 $ha^{-1}year^{-1}$)	C credit (I) (AUD ha^{-1} $year^{-1}$)		N input to stabilise C storage[c] (kg N $ha^{-1}year^{-1}$)	N cost to stabilise C storage[d] (II) (AUD $ha^{-1}year^{-1}$)	Financial returns (I–II) (AUD $ha^{-1}year^{-1}$)	
	Mean	95% CI		Voluntary[a]	CFI[b]			Voluntary	CFI
Conservation tillage									
0–10	150	97 to 210	551.5	1.8	12.7	15.0	19.6	−17.8	−6.9
11–20	100	58 to 142	364.8	1.2	8.4	10.0	13.0	−11.9	−4.6
21–30	40	1 to 86	147.8	0.5	3.4	4.0	5.3	−4.8	−1.9
31–40	6	−6 to 20	0	0	0	NA	NA	NA	NA
Residue retention									
0–10	147	40 to 276	537.5	1.7	12.4	14.7	19.1	−17.5	−6.8
11–20	71	46 to 98	260.3	0.8	6.0	7.1	9.3	−8.4	−3.3
21–30	21	−15 to 65	0	0	0	NA	NA	NA	NA
31–40	4	−4 to 13	0	0	0	NA	NA	NA	NA
Use of pasture									
0–10	132	42 to 259	482.2	1.5	11.1	13.2	17.2	−15.7	−6.1

11–20	NA	NA	NA	NA	NA	NA	NA	NA	NA
21–30	75	57 to 95	275.4	0.9	6.3	7.5	9.8	−8.9	−3.5
31–40	60	40 to 79	218.9	0.7	5.0	6.0	7.8	−7.1	−2.8
Fertiliser N application									
0–10	67	20 to 112	247.1	0.8	5.7	6.7	8.8	−8.0	−3.1
11–20	40	18 to 60	145.2	0.5	3.3	4.0	5.2	−4.8	−1.8
21–30	NA	NA	NA	NA	NA	NA	NA	NA	NA
31–40	14	7 to 22	51.7	0.2	1.2	1.4	1.8	−1.7	−0.7

[a]calculated for a carbon price of USD 3.35 (AUD 3.19) Mg^{-1} CO_2-equivalent in the voluntary carbon market[32], based on exchange rate of AUD: USD=1:1.05.

[b]calculated for a Kyoto Australian carbon credit unit of AUD 23 Mg^{-1} CO_2-equivalent issued under the Carbon Farming Initiative (CFI) in Australia[10].

[c]estimated by soil C:N ratio of 10:1 (ref. 8).

[d]estimated by N cost of AUD 600 Mg^{-1} urea (46%N)[33].

Economic Assessment

The relative soil C gain represents a carbon credit of AUD 1.8 ha^{-1} $year^{-1}$ by practising conservation tillage, AUD 1.7 ha^{-1} $year^{-1}$ for residue retention, AUD 1.5 ha^{-1} $year^{-1}$ for the use of pasture, and AUD 0.8 ha^{-1} $year^{-1}$ for fertiliser N application, during the first 10 years of a management change (Table 2). When the cost of N to stabilise C storage was considered, the financial loss based on the C price in the voluntary carbon market was AUD 17.8 ha^{-1} $year^{-1}$ for conservation tillage, AUD 17.5 ha^{-1} $year^{-1}$ for residue retention, AUD 15.7 ha^{-1} $year^{-1}$ for pasture use, and AUD 8.0 ha^{-1} $year^{-1}$ for fertiliser N application (Table 2). Based on the CFI, the financial loss was reduced to AUD 6.9, 6.8, 6.1 and 3.1 ha^{-1} $year^{-1}$ correspondingly (Table 2).

DISCUSSION

Based on the available field studies that covered a wide range of soil and climatic conditions across Australia, our results suggest that soil at 0–10 cm depth is the most responsive to the targeted management practices in Australia's cropped soils. Increases in soil C as a result of these management practices have been attributed to enhanced biomass (above- and below-ground) production, litterfall and hence return of residues to the soil, and also to improved soil aggregation that protects the C compounds from rapid decomposition[4, 6, and 12]. However, the small increases (7–13%) in the C concentration in the surface layer of Australian agricultural soils may not be technically credible due to sampling errors, spatial variability and uncertainties associated with measurement and analysis[13]. The positive management effects on soil C concentration or C stock were not detectable in soil deeper than 10 cm (Fig. 1; Table 1). This result is not unique to Australian cropping systems. For example, in studies comparing conservation tillage with conventional tillage in other countries, it has been shown that soil C may increase only in the surface 10 cm of the soil, and remain unchanged or even decrease in deeper layers[7, 14].

Carbon accumulation can be increased in the surface 0–10 cm of soil, but that is also the layer from which it can be most readily lost from Australian agricultural soils[5], and the management-induced increases of soil C in this surface layer are very vulnerable to environmental and management pressures[9]. It is therefore necessary to examine whether an agricultural practice sustains surface soil C over time. This is important because the intent of Australia's CFI is to encourage farmers and land managers to use an approved methodology that enables permanent storage of soil C (100+ years) [10]. However, we found that the relative soil C gain in Australian cropping systems decreased with time, regardless of agricultural practices (Table 2). Moreover, soil organic C in many Australian cropping soils is not in state of C equilibrium, and the soils are continuing to lose C[15]. The observed relative C gain under improved management treatments might therefore have resulted from a reduced or discontinued soil C loss instead of real C accrual[15].

In terms of sustainability, the use of pasture outperformed the other practices after 31–40 years (Table 2). This highlights the potential use of pasture in C sequestration in Australian agriculture, for at least 40 years. However, at least three problems must be considered with increased pasture use, especially permanent pasture, in Australia: (i) increased methane emissions for ruminant animals grazing those pastures, (ii) the extra demand for irrigation water to maintain those pastures (if they are to be effective in storing more soil C), and (iii) the less efficient use of solar radiation per hectare in producing food from animals compared to food from crops.

When the cost of N to stabilise C accrual was considered, we estimated that none of the four management practices resulted in any net profit based on the recent C trading price of AUD 3.19 Mg^{-1} CO_2-e in the voluntary market, but a loss of AUD 8–18 ha^{-1} $year^{-1}$ for the first 10 years of implementation of the practice (Table 2). The loss was reduced to AUD 3–7 ha^{-1} $year^{-1}$ under the CFI in Australia (currently at AUD 23 Mg^{-1} CO_2-e). The minimum C price required for a net profit was estimated to be AUD 36 Mg^{-1} CO_2-e. These results indicate that current agricultural management options and the current value of conserved C are not economically viable. More importantly, for a carbon trading scheme to be credible, change in soil C has to be verifiably measured or modelled cost-effectively at the farm scale[13].

In terms of benefits to the environment, based on an area of approximately 100 M ha of managed cropland and modified pasture in Australia in 2006 (ref. 16), we estimate that the gain in CO_2-e equates to 4.6–10.3% of total national greenhouse gas (GHG) emissions in the same year of 533.1 Tg CO_2-e[17]. This estimate assumes 100 percent take-up of the improved practices and does not consider the potential impact of these practices on emissions of methane from grazing animals and nitrous oxide from additional N fertiliser use[18, 19, 20, and 21]. While the effects of improved management practices on soil C stock vary with temperature and rainfall[5], scientific knowledge of the capacity of the soil to store C under future climates, such as warmer and drier conditions and higher atmospheric CO_2 levels[22, 23, 24], is insufficient for Australian cropping systems.

Our analysis shows that improved management practices (conservation tillage, residue retention, the use of pasture or fertiliser N application) increased soil C by 0.05–0.15 Mg C ha^{-1} $year^{-1}$ only in the surface 10 cm of Australian croplands. The effects were more pronounced during the first 10 years of implementation of improved management practices.

Because of the technical uncertainty in measuring small soil C changes, and the higher cost of stabilising any gain in soil C compared with the C credit earned, carbon sequestration in Australian agricultural soils as a means of offsetting a significant part of the national GHG emissions is technically limited and economically nonviable at the present time. These results have major implications for improvement in the measurement and modelling of soil C changes at the field scale, and enactment of carbon reduction legislations in Australia and other parts of the world.

METHODS

Database Compilation

This meta-analysis is based on studies (see Supplementary Notes online) of the effects of agricultural land management practices (conservation tillage, residue retention, the use of pasture, and N fertilization) on soil C concentration (%C) and C stock (kg C ha^{-1}) in Australian cropping systems published between 1984 and December 2012. We performed extensive keyword searches of several databases (Web of Science, CAB Abstracts, Scopus, Agricola and Agriculture and Natural Resources Index), and by examining the reference lists of cited references. The search terms are listed in Supplementary Notes online. Original data were extracted from tables whereas data values presented in figures were obtained using digitizing software.

The following criteria were set for a study to be included in the analysis. First, the sample sizes and means of soil C concentrations or stocks had to be reported for control and treatment plots. Second, treatments were applied at the same farm site throughout the experimental period, i.e. paired-site studies or surveys were excluded. Third, the study must have been conducted under field conditions for at least three consecutive years. For multiple observations at any one site over time, individual observations were selected at intervals of at least every third year. The duration of an experiment was considered as the number of years after the commencement of a treatment. Conservation tillage encompassed no-till and reduced tillage. Residue retention comprised the incorporation of residues and surface retention. The use of pasture included conversion of cropped land to pasture and the incorporation of pasture into a rotation.

Because the soil C stock is an amount of C in a mass of soil in a defined area (1 ha) to a defined depth (commonly 30 cm), C concentration must be multiplied by the dry bulk density (BD) to obtain the soil C stock (kg ha^{-1}). Hence when not reported, soil C stock was calculated by multiplying the %C values for each soil depth interval by its BD and the depth increment. In experiments where BD was not provided, a pedotransfer function was used to estimate BD. For the studies included in this meta-analysis, we considered two published pedotransfer functions for the estimation of BD by comparing the regression between measured and estimated BD values. The two functions were $BD=1.66-0.308\,(OC)^{0.5}$ (equation (1)[25]) and $BD=1.398-0.0047\,(\%Clay)-0.042\,(\%OC)$ (equation (2)[26]), where the BD units were Mg m^{-3} and OC=organic carbon (%). Based on equation (2), the regression coefficient between measured and estimated BD values was 0.64 ($p < 0.001$), whereas that based on equation (1) was 0.32 ($p < 0.001$). We therefore based our estimation of BD on equation (2).

Ideally, the soil C stock for treatments and controls should be compared on an equivalent mass, rather than a fixed depth basis, because of the changes in BD that may occur under different management practices[27]. In our study, soil depth was not adjusted to account for changes, if any, in BD associated with changes in management practices unless the authors of the study had already done so. Not adjusting soil depth may in some cases result in a small error in the estimate of the management effects on soil C stocks per unit land area[28].

The database of the meta-analyses is in Supplementary Tables S1–S4 online. The dataset was compiled from 56 studies with 435 direct comparisons, including 172 comparisons for the effects of conservation tillage, 116 for residue retention, 83 for the use of pasture and 64 for fertiliser N application. The potential patterns of variation in the effects of various management practices were also assessed by including categorical variables in the meta-analysis models. These variables included experimental duration (0–10, 11–20, 21–30 and 31–40 years) and soil depth (0–10, 10–20, 20–30 and 30–40 cm). Soil sampling depths varied between studies. For studies in which the sampling depths did not match the defined categories, each data point for C concentration was categorised as the average depth of the soil layer from which it was collected[14]. Soil C stock data were included only if the sampling depth matched the defined categories.

Meta-Analysis

We used the natural log of the response ratio ($r = \overline{x}_t / \overline{x}_c$, where $\overline{x}_t$ and $\overline{x}_c$ are the means of the treatment and control groups, respectively) [29] as a metric for the analysis of treatment effects on soil C concentration. These results are reported as the percentage change under treatment effects ($(r - 1) \times 100$). Negative percentage changes indicate a decrease in soil C compared with the control and positive values indicate an increase due to an improved management practice. The effect of management practices on the rate of change in the soil C stock relative to the control was expressed as U (kg C ha^{-1} year^{-1})=(amount of C in treatment plot – amount of C in control plot)/experimental duration.

Effect sizes reported in previous meta-analyses on soil C storage were weighted by the inverse of the pooled variance[6], replication[23] or unweighted[2]. The studies collected in our database did not always include published variances. In addition, the variance-based weighting function could result in extreme weights for some studies, while weighting by replication could yield less extreme weights[30]. Therefore our analysis was performed on effect sizes weighted by a function of sample size using:

$$weight = (n_c \times n_t) / (n_c + n_t),$$

Where n_c and n_t are the number of replicates of the control and treatment groups, respectively[30].

Mean effect sizes and 95% confidence intervals were generated by bootstrapping (4,999 iterations) using MetaWin 2.1[31]. A fixed-effects model or a mixed-effects model is technically not applicable for non-parametric meta-analytic procedures based on weighting by replication. However, to perform correct bootstrapping using Meta Win a fixed-effects model had to be selected. The effects of agricultural management practices were considered significant if the confidence intervals did not overlap with zero.

Economic Assessment

We performed an economic assessment of the improved management practices based on the C price of USD 3.35 (AUD 3.19) Mg^{-1} CO_2-e in the voluntary carbon market in 2008 (ref. 32) or Kyoto Australian carbon credit unit at AUD 23 Mg^{-1} CO_2-e in 2013 (ref. 10), a C: N ratio of 10:1 across agricultural soils[8], and the cost of urea (46% N) at AUD 600 Mg^{-1} (ref. 33). These estimates exclude any increase in N fertiliser cost passed on to the landholder, which is incurred by the manufacturer having to pay the mandatory C price (currently AUD 23 Mg^{-1} CO_2-e[10]) associated with GHG emissions during production. The costs of other nutrients such as phosphorus and sulphur to match C gain are not included, nor did any of the compliance costs such as reporting and auditing associate with participation in the CFI.

ACKNOWLEDGEMENTS

We wish to thank Emeritus Professor Robert White and Dr Ian Willett for valuable comments on our manuscript, Dr. Kees Jan van Groenigen for assistance in meta-analytic procedures, and the authors who provided information for the construction of the database. This work was supported by Australian Research Council, Department of Agriculture, Fisheries and Forestry and Dean's strategic research fund of our School.

AUTHOR CONTRIBUTIONS

R.R. and D.C. initiated the concept of the study. R.R., D.C., A.R.M. and S.K.L. designed the investigation. S.K.L. and D.C. contributed to data extraction from the literature and the construction of the database. S.K.L., D.C. and A.R.M. performed the statistical analyses. All authors were involved in writing the paper.

REFERENCES

1. Stockmann, U. *et al.* The knowns, known unknowns and unknowns of sequestration of soil organic carbon. *Agric. Ecosyst. Environ.* 164, 80–99 (2013).
2. Guo, L. B. & Gifford, R. M. Soil carbon stocks and land use change: a meta-analysis. *Glob. Change Biol.* 8, 345–360 (2002).
3. Lal, R. Soil carbon sequestration impacts on global climate change and food security.*Science* 304, 1623–1627 (2004).
4. Conant, R. T., Paustian, K. & Elliott, E. T. Grassland management and conversion into grassland: effects on soil carbon. *Ecol. Appl.* 11, 343–355 (2001).
5. Luo, Z., Wang, E. & Sun, O. J. Soil carbon change and its responses to agricultural practices in Australian agro-ecosystems: a review and synthesis. *Geoderma* 155, 211–223 (2010).
6. Lu, M. *et al.* Minor stimulation of soil carbon storage by nitrogen addition: a meta-analysis.*Agric. Ecosyst. Environ.* 140, 234–244 (2011).
7. Luo, Z., Wang, E. & Sun, O. J. Can no-tillage stimulate carbon sequestration in agricultural soils? A meta-analysis of paired experiments. *Agric. Ecosyst. Environ.* 139, 224–231 (2010).
8. Carlyle, J. C., Charmley, E., Baldock, J. A., Polglase, P. J. & Keating, B. in *Adapting Agriculture to Climate Change: Preparing Australian Agriculture, Forestry and Fisheries for the Future* (eds Stokes C.,& Howden M., eds.) 229–244 (CSIRO Publishing, 2010).
9. arquharson, R. & Baldock, J. *Soil Carbon Sequestration Potential: a Review for Australian Agriculture* (CSIRO Publishing, 2010).

10. Department of Climate Change and Energy Efficiency. . *The Carbon Farming Initiative Handbook* Version 1.0 (Australian Department of Climate Change and Energy Efficiency, 2012), accessed 10 June 2013 at

11. Gruber, N. & Galloway, J. N. An earth-system perspective of the global nitrogen cycle. *Nature*45, 293–296 (2008).

12. Six, J., Paustian, K., Elliott, E. T. & Combrink, C. Soil structure and organic matter: I. Distribution of aggregate-size classes and aggregate-associated carbon. *Soil Sci. Soc. Am. J.*64, 681–689 (2000).

13. Walcott, J., Bruce, S. & Sims, J. *Soil Carbon for Carbon Sequestration and Trading: a Review of Issues for Agriculture and Forestry* (Australian Bureau of Rural Sciences, 2009).

14. Angers, D. A. & Eriksen-Hamel, N. S. Full-inversion tillage and organic carbon distribution in soil profiles: a meta-analysis. *Soil Sci. Soc. Am. J.* 72, 1370–1374 (2008)

15. Sanderman, J. & Baldock, J. A. Accounting for soil carbon sequestration in national inventories: a soil scientist's perspective. *Environ. Res. Lett.* 5, 034003 (2010).

16. Bureau of Rural Sciences. . *Land Use of Australia* Version 4, 2005–06 data (Australian Bureau of Rural Sciences, 2010).

17. Department of Climate Change and Energy Efficiency. . *National Inventory Report 2009 Australian National Greenhouse Accounts* Volume 3 (Australian Department of Climate Change and Energy Efficiency, 2011).

18. Barclay, A. Farming carbon. *ECOS* 156, 24–26 (CSIRO Publishing 2010).

19. Liu, L. & Greaver, T. L. A review of nitrogen enrichment effects on three biogenic GHGs: the CO_2 sink may be largely offset by stimulated N_2O and CH_4 emission. *Ecol. Lett.* 12, 1103–1117 (2009).

20. Wang, W., Dalal, R. C., Reeves, S. H., Butterbach-Bahl, K. & Kiese, R. Greenhouse gas fluxes from an Australian subtropical cropland under long-term contrasting management regimes. *Glob. Change Biol.* 17, 3089–3101 (2011).

21. Van Kessel, C. *et al.* Climate, duration, and N placement determine N_2O emissions in reduced tillage systems: a meta-analysis. *Glob. Change Biol.* 19, 33–44 (2012).

22. Conant, R. T. *et al.* Temperature and soil organic matter decomposition rates – synthesis of current knowledge and a way forward. *Glob. Change Biol.* 17, 3392–3404 (2011).

23. Hungate, B. A. *et al.* Assessing the effect of elevated carbon dioxide on soil carbon: a comparison of four meta-analyses. *Glob. Change Biol.* 15, 2020–2034 (2009).

24. Schils, R. *et al. Review of Existing Information on the Interrelations between Soil and Climate Change (ClimSoil). Final Report* (European Commission, 2008).

25. Alexander, E. B. Bulk densities of California soils in relation to other soil properties. *Soil Sci. Soc. Am. J.* 44, 689–692 (1980).

26. Bernoux, M., Arrouays, D., Cerri, C., Volkoff, B. & Jolivet, C. Bulk densities of Brazilian Amazon soils related to other soil properties. *Soil Sci. Soc. Am. J.* 62, 743–749 (1988).

27. Powlson, D. S., Whitmore, A. P. & Goulding, K. W. T. Soil carbon sequestration to mitigate climate change: a critical re-examination to identify the true and the false. *Eur. J. Soil Sci.* 62, 42–55 (2011).

28. Post, W. M. & Kwon, K. C. Soil carbon sequestration and land-use change: processes and potential. *Glob. Change Biol.* 6, 317–327 (2000).

29. Hedges, L. V., Gurevitch, J. & Curtis, P. S. The meta-analysis of response ratios in experimental ecology. *Ecology* 80, 1150–1156 (1999).

30. Van Groenigen, K. J., Osenberg, C. W. & Hungate, B. A. Increased soil emissions of potent greenhouse gases under increased atmospheric CO_2. *Nature* 475, 214–216 (2011).

31. Rosenberg, M. S., Adams, D. C. & Gurevitch, J. *MetaWin: Statistical Software for Meta-Analysis* Version 2 (Sinauer Associates Inc., 2000).

32. Hamilton, K., Sjardin, M., Shapiro, A. & Marcello, T. *Fortifying the Foundation: State of the Voluntary Carbon Markets 2009.* (Ecosystem Marketplace and New Carbon Finance, 2009), accessed 28 December 2012 at

33. Mulvany, J. *Managing Your Farm Business in a Variable Climate, Riding the Diary Roller Coaster Forum* (Australian Department of Agriculture, Forestry & Fisheries and Dairy Australia, 2011), accessed 19 November 2012 at.

Chapter 2

Higher Levels of Multiple Ecosystem Services are found in Forests with More Tree Species

Lars Gamfeldt[1, 2], Tord Snäll1, Robert Bagchi[3], Micael Jonsson[4], Lena Gustafsson[1], Petter Kjellander[5], María C. Ruiz-Jaen[6], Mats Fröberg[7, 8], Johan Stendahl[8], Christopher D. Philipson[9], Grzegorz Mikusiński[5], Erik Andersson[10, 11], Bertil Westerlund[1, 2], Henrik Andrén[5], Fredrik Moberg[1, 1], Jon Moen[4], and Jan Bengtsson[1]

[1]Department of Ecology, Swedish University of Agricultural Sciences, Box 7044, SE-75007 Uppsala, Sweden

[2]Department of Biological and Environmental Sciences, University of Gothenburg, Box 461, SE-40530 Gothenburg, Sweden

[3]Department of Biological and Biomedical Sciences, Durham University, South Road, Durham DH1 3LE, UK

[4]Department of Ecology and Environmental Science, Umeå University, SE-90187 Umeå, Sweden

[5]Grimsö Wildlife Research Station, Department of Ecology, Swedish University of Agricultural Sciences, SE-730 91 Riddarhyttan, Sweden

[6]Environmental Change Institute, South Parks Road, Oxford OX1 3QY, UK

[7]Department of Aquatic Sciences and Assessment, Swedish University of Agricultural Sciences, Box 7082, SE-75007 Uppsala, Sweden

[8]Department of Soil and Environment, Swedish University of Agricultural Sciences, Box 7014, SE-75007 Uppsala, Sweden

[9]Institute of Evolutionary Biology and Environmental Studies, University of Zurich, Winterhurerstrasse 190, 8057 Zurich, Switzerland

[10]Southern Swedish Forest Research Centre, Swedish University of Agricultural Sciences, Box 49, SE-23053 Alnarp, Sweden

[11]Stockholm Resilience Centre, Stockholm University, SE-10691 Stockholm, Sweden

[12]Department of Forest Resource Management, Swedish University of Agricultural Sciences, Box 7001, SE-90183 Umeå, Sweden

ABSTRACT

Forests are of major importance to human society, contributing several crucial ecosystem services. Biodiversity is suggested to positively influence multiple services but evidence from natural systems at scales relevant to management is scarce. Here, across a scale of 400,000 km^2, we report that tree species richness in production forests shows positive to positively hump-shaped relationships with multiple ecosystem services. These include production of tree biomass, soil carbon storage, berry production and game production potential. For example, biomass production was approximately 50% greater with five than with one tree species. In addition, we show positive relationships between tree species richness and proxies for other biodiversity components. Importantly, no single tree species was able to promote all services, and some services were negatively correlated to each other. Management of production forests will therefore benefit from considering multiple tree species to sustain the full range of benefits that the society obtains from forests.

INTRODUCTION

Ecosystems provide a wealth of benefits to human society, and the provision of such ecosystem services depends fundamentally on functions performed by organisms[1]. This has led scientists to enquire how the diversity and composition of communities may regulate ecosystem functions[2, 3]. A large body of evidence has established that species diversity promotes ecosystem functions under experimental conditions[4, 5]. There are, however, many exceptions to the positive diversity–function relationship[6]. In addition, most experiments have been conducted at limited temporal or spatial scales (but see refs 7, 8). It is thus uncertain if conclusions based on results from these studies can be extended to the scales relevant to policy makers.

Thus far, studies that manipulated biodiversity and monitored the consequences mainly focused on ecosystem functions, but because functions have complex links to ecosystem services[1, 9], the policy implications of these studies are unclear[5]. There are relatively few large-scale observational studies explicitly aimed at quantifying the importance of biodiversity, and these have mostly analysed single-ecosystem functions or services[10, 11, 12]. Furthermore, while there are multiple functions that regulate services from ecosystems, few studies have investigated the role of biodiversity for multiple ecosystem functions jointly (but see, refs 13, 14, 15, 16), and none of these have focused on services *per se*. In a recent comprehensive review, a majority of the included ecosystem services were related to biodiversity in the direction expected from predictions[3]. However, for many of the studied services, the evidence for beneficial effects of biodiversity was mixed, or there were not enough data for a thorough evaluation[3]. Studies that explicitly investigate the link between biodiversity and multiple ecosystem services across regional scales are thus urgently needed if research is to be informative for management and policy, such as the Intergovernmental Platform on Biodiversity and Ecosystem Services[17].

We examined the relationships between multiple ecosystem services and both tree species richness and tree biomass in boreal and temperate forest. These extensive biomes are of high global importance[1], constituting around 27% (8.7 million km^2) and 16% (5.2 million km^2) of the world's forests, respectively[18]. The majority of studies on the importance of tree diversity have focused on productivity, and there is evidence for both non-significant[19] and positive effects[20]. We studied production forests, that is, forests that are subjected to a silvicultural system of harvesting and planting or natural regeneration from seed trees. Production forest is the dominant system in the sampled region, which covered an area of 400,000 km^2 and a span of 13.7 degrees of latitude. These forests are relatively species poor. The maximum number of tree species in any one sampling plot in our study was 10, and only 1.5% of the plots hosted more than five species. Six ecosystem services[1] were included: tree biomass production, topsoil carbon storage, berry production, game production potential, understory plant species richness and dead wood occurrence. Production of tree biomass is a major global industry, contributing about US$ 400 billion to the annual gross world product in terms of wood[1]. Forests and their soils store about 45%of the terrestrial carbon, and act as a crucial sink for anthropogenic carbon emissions[21]. Both production of berries and game are of large monetary and recreational value in many countries[1]. Dead wood occurrence represents a supporting function for other ecosystem services, as it affects ecological processes such as nutrient cycling and soil formation[1, 22], and it also indicates several other components of biodiversity. Understory plant species richness differs somewhat from the other five ecosystem services, as it is not linked to biological processes or biological production, but instead can be classified as a cultural ecosystem service that provides aesthetic, educational and recreational value to human beings[1]. Alternatively, and with recent terminology, understory plant species richness may be viewed as having a value of its own[23]. Safeguarding biodiversity at all levels of organization is also a central goal of the Convention on Biological Diversity.

RESULTS

Relationships between Tree Species Richness and the Services

The relationships between tree species richness and tree biomass production, berry production (of bilberries) and game production potential were positively hump-shaped (Fig. 1a), taking into account important environmental variables, such as climate, soil nutrients and forest age (Methods). On average, biomass production at the mean forest age was 54% greater with five tree species than with only one species (Fig. 1a). Corresponding percentages for berry production and game production potential were 45% and 20%, respectively. Soil carbon storage and understory plant species richness increased with tree species richness (Fig. 1b). Soil carbon storage was 11%greater, and understory plant species richness 31% greater, with five than with one tree species. The occurrence of dead wood increased at high tree species richness (Fig. 1f). An alternative way of evaluating how productivity changes with tree species richness is to quantify the probability that a given species richness would have higher productivity than a monoculture. In our analyses, tree biomass production is 41% more likely to be higher on plots with five species than on plots with one species. Excluding the effect of the biomass of individual tree species from the models did not change the positive relationships between the services and tree species.

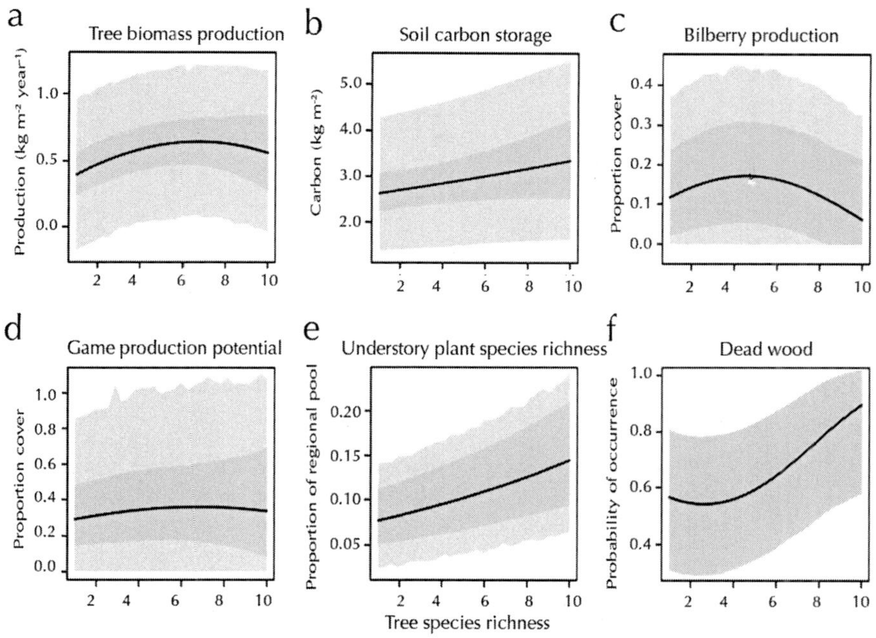

Figure 1: Relationships between tree species richness and ecosystem services. (a) Tree biomass production; (b) soil carbon storage; (c) bilberry production; (d) game production potential; (e) understory plant species richness; (f) occurrence of dead wood. We show mean relationships (black) and 95% Bayesian confidence intervals for the relationships excluding (dark green) and including the residual variation (light green). Other model explanatory variables were kept at mean levels.

Individual Tree Species and Services

None of the six ecosystem services was consistently explained by the biomass of the same tree species (Fig. 2). For example, there were strong positive relationships between the biomass of spruce and tree biomass production and occurrence of dead wood, while pine biomass was positively related to bilberry production, and birch biomass to soil carbon storage. Beech was not positively related to any service.

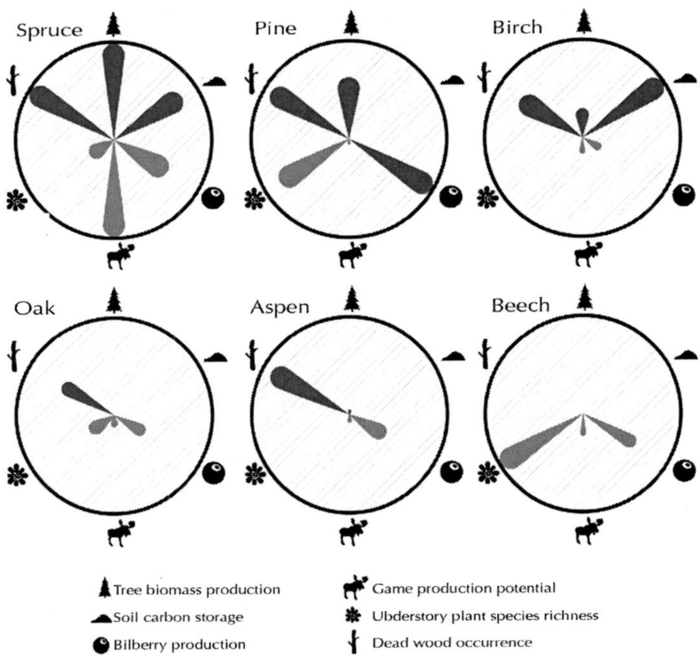

Figure 2: Strength of the relationships between tree biomasses and ecosystem services. The lengths of the 'petals' in these flower diagrams (one diagram for each tree species) reflect transformed effect sizes for the relationships in the main models. Specifically, the effect sizes (β_n in equation (1)) have been transformed to range between 0 and 1, so that for each ecosystem service the effect size for the tree species with the greatest effect equals the absolute value of 1. Blue are positive relationships, red are negative relationships, and the outer circles represent the maximum value of |1|. The six ecosystem services are labelled by their respective symbols.

Relationships between Ecosystem Services

Some of the services were positively related to each other (Fig. 3), but we also found some notable trade-offs. First, there were trade-offs between tree biomass production and occurrence of dead wood, game production potential and bilberry production. Second, dead wood occurrence and game production potential were negatively correlated.

Figure 3: Pairwise relationships between ecosystem services. Blue are positive and red are negative relationships. Grey reflects relationships with no established inter-relationship. The six ecosystem services are labelled by their respective symbols.

DISCUSSION

The pros and cons of species mixtures for productivity and other ecosystem functions have been discussed at length since the early 19th century[2, 24, 25]. Not until recently, however, have scientists begun to explicitly investigate how species diversity might be important for the simultaneous provision of multiple functions or services[13, 14, 15, 26, 27]. Our results from boreal and temperate production forests show that the relationships between tree species richness and multiple ecosystem services were positive to positively hump-shaped, and that all services attained higher levels with five tree species than with one species. Although the relatively high level of tree biomass production with five compared with one tree species may seem both impressive and surprising, we note that similar effect sizes have been found previously[11]. Another observational study, in which climatic and environmental conditions were controlled for, also found a strong and positive effect of tree diversity on productivity[12]. Because controlled field studies on mixed-stand effects have revealed both non-significant and positive effects of diversity[19, 20], future research should aim to understand and reconcile results from controlled experiments and observational studies. There are, unfortunately, no previous studies on the relationships between tree species richness and the other five services that allow for comparisons with our findings.

We found little support for a general importance of the biomass of particular species in explaining all services (Fig. 2). The finding that different species were most strongly related to different services indicates that monoculture practices will lead to reduced provision of at least some of the services we considered. For example, among all tree species considered, only pine was strongly and positively related to bilberry production (see also ref. 28) and birch showed the strongest positive relationship with soil carbon storage. As such, our results suggest the importance of tree species mixtures for the continued provisioning of ecosystem services from the 2-billion hectares of forest in the world currently managed as production forests or used for multiple purposes (55% of all forests)[29].

The results also indicate trade-offs between some services (Fig. 3). The trade-off between tree biomass production and dead wood occurrence is probably an effect of stand age; tree production generally decreases with age, whereas the opposite is true for occurrence of dead wood[30]. The negative correlation between dead wood occurrence and game production potential is most likely also partly explained by forest stand age; herbivore forage plants are more common in young forests, whereas dead wood more commonly occurs in old forests. Negative relationships between tree biomass production and the production of both bilberry and food for game are likely consequences of light limitation and lower precipitation through-fall at the ground level in closed productive forests[31]. Unlike previous findings[32], we thus found that some ecosystem services may come at the cost of others. Identifying such trade-offs is highly relevant for policy, enabling users of forest ecosystems to understand and balance the pros and cons of different management scenarios[33]. Although the trade-offs we found imply that it will be difficult to maximize all ecosystem services simultaneously at the stand scale, the positive relationships between tree diversity and individual services (Fig. 1) suggest that adjacently located monocultures would not optimize the provision of ecosystem services at the landscape scale. Instead, adjacent stands, each with multiple species but in different combinations, might be the best way to provide multiple ecosystem services at the landscape scale.

The negligible effects of the biomass of single tree species on the relationship between tree species richness and ecosystem services in our study suggest complementarity among tree species, for example, resource partitioning, often referred to as 'complementarity effects' (or effects of species richness *per se*) in experimental studies[5]. Interactions between plants and associated microbiota, and higher resistance of a diverse community to disease, insects or pathogens may also be factors explaining the observed relationships between some of the services and tree species richness[34].

Additional environmental conditions may directly or indirectly explain some of the unexplained variation in the services[34, 35]. For example, nitrogen availability (not measured in the Swedish National Forest Inventory) may influence forest productivity. However, the ratio of total carbon to total nitrogen in the topsoil that we used is a good indicator of nitrogen availability[36]. Also, non-measured environmental variables that are correlated with altitude or latitude could be important; for example, plots in the mountain region (in northwest Sweden) generally have fewer tree species. However, we accounted for temperature and humidity, which are the most important climate variables and correlated with altitude and latitude. Moreover, maximum tree species richness was relatively evenly distributed across the sampling region.

While our analyses were on the regional scale, there certainly may be differences at the local scale in the strength of the relationships between tree species richness and ecosystem services. All services may thus not be maximized similarly within landscapes across the whole sample region. Furthermore, we focused on the current distribution and composition of forests, but future changes in the environment may affect ecosystem services both directly and indirectly[37]. Global change may thus result in society not having all the management options currently available. Scenarios of changes in forest species composition and richness due to global change must be considered in analyses of future provisioning of ecosystem services[38].

The majority of previous efforts to disentangle the effects of tree species richness on ecosystem functions or services in forests have focused on productivity of woody biomass[11, 12, 39, 40]. Mixed forest stands can indeed be more productive than monocultures[20], and several studies have found large-scale positive relationships between tree diversity and wood production[11, 12, 41]. Positive relationships between biodiversity and individual ecosystem functions or services have also been observed in aquatic realms, such as the open ocean[42], the deep sea[43], coral reefs[44] and lakes[45]. Our study, which accounts for important environmental conditions, shows consistent positive relationships between tree species richness (contrasting plots with five and one tree species) and multiple ecosystem services. It also highlights the importance of conserving a variation of tree species, to safeguard a future potential of high levels of multiple ecosystem services. As such, our research expands on previous work[16, 46], none of which considers multiple services. Our study revealed complementarity among species; no single species can sustain multiple services at high levels simultaneously. This is particularly pertinent at a time when demands on ecosystems to provide humans with several important services are increasing[47] at the same time as global change and human expansion is threatening the integrity of ecosystems, their biodiversity and functioning[1,3, 44, 48]. Our results strongly support the integration of production and conservation in natural resource management, and align with the 'ecosystem approach', the primary framework for action under the Convention of Biodiversity[49]. We show that moving towards multi-species management can better realize the full potential of several economically, ecologically and culturally valuable ecosystem services.

METHODS

Description of the Data

We used a nationwide forest data set from the Swedish National Forest Inventory and the Swedish Survey of Forest Soils and Vegetation. The inventory uses a randomly planned regular sampling grid[50], and includes around 4,500 permanent tracts with each tract being surveyed once every 5 years. The tracts, which are rectangular in shape and are of different dimensions in different parts of the country, consists of 8 (in the north) to 4 (in the south) circular sample plots (each plot 314 m^2). From the Inventory database, we extracted data from 1999 to 2002. We used only plots on 'productive forest' (average production of standing volume, stem volume over bark >1 $m^3\ ha^{-1}\ year^{-1}$), which had not been harvested, cleared or thinned during the previous 5 years before the survey. The previous time period of 5 years was also used for calculating biomass production. To be included in our analyses, the plots had to be located on only forest land, for example, not including any river, road, grassland and so on. We excluded plots where the biomass production of any of the focal tree species was greater than the 99% quantile (which is the standard procedure in analyses of data from the Swedish National Forest Inventory). This procedure excluded plots with unrealistically high values of production. Our selection resulted in a data set with a total of 4,335 sample plots distributed across 1,401 tracts. We included explanatory variables hypothesized to explain the variation in the six modelled ecosystem services, and squared variables and interactions, whenever biologically reasonable .

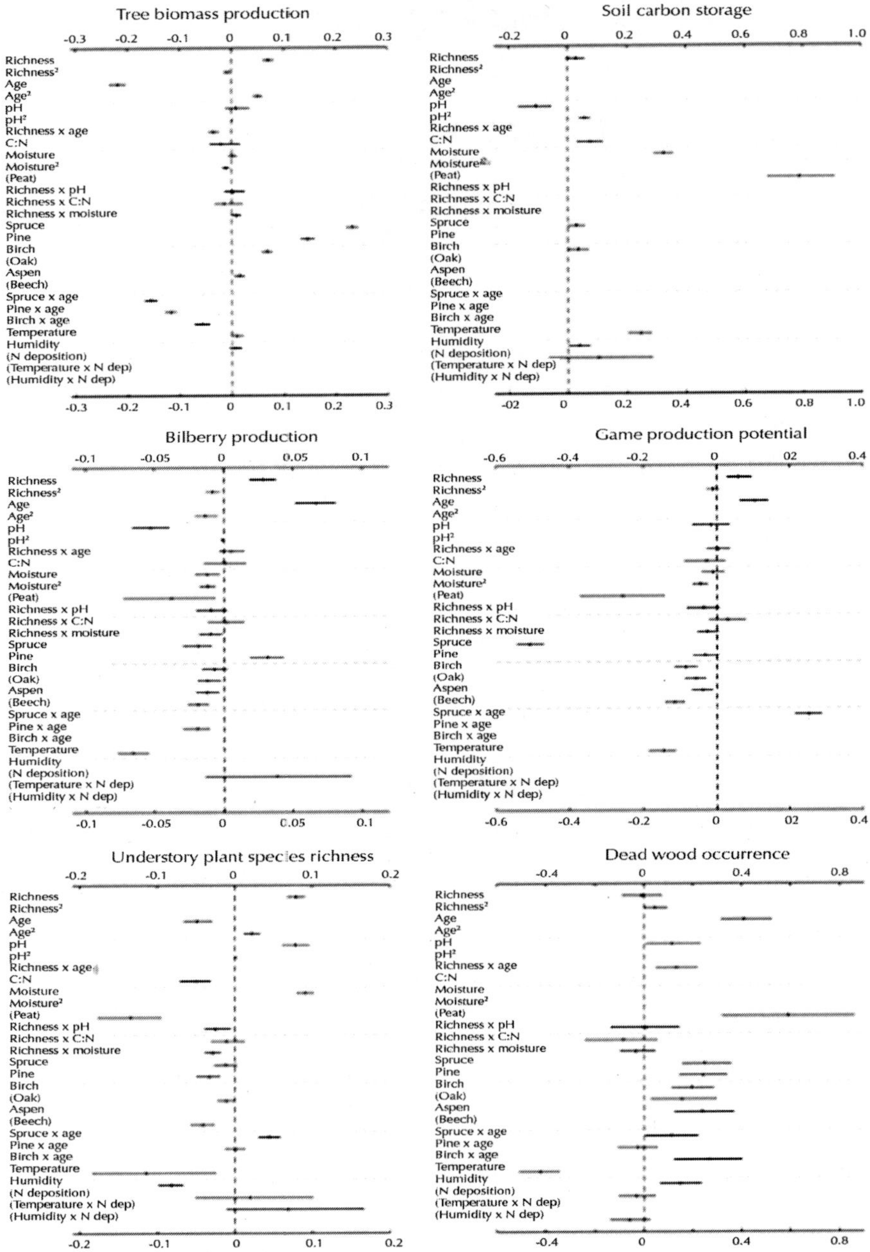

Figure 4: Parameter estimates for the main model for each ecosystem service. The modes (vertical lines) and the 95% Bayesian confidence intervals (thin horizontal lines) for the parameters (β_n in equation (1) and ρ_m in equation (3)) are shown for each service (tree biomass production, soil carbon storage, bil-

berry production, game production potential, understory plant species richness and occurrence of dead wood). Variables or interactions in parentheses were not included in the final main model. Dashed lines are for visual aid.

Statistical Modelling

To model the six ecosystem services as a function of tree species richness, we used hierarchical Bayesian-generalized linear models[51]. The Bayesian approach is convenient for fitting complex hierarchical models with a formal mathematical treatment of the natural variability and data uncertainty, for example, when the recording of the explanatory variables is incomplete.

For each ecosystem service, we fitted the following models: the main model, on which our results are based unless otherwise mentioned (Figs 1 and 2); a model aimed to investigate how the tree species biomasses affected the relationship between tree species richness and the services, specifically the main model but excluding the tree biomass variables; five models aimed to investigate the pairwise relationships between the ecosystem services, excluding all other explanatory variables (Fig. 3). In the following, we exemplify the modelling using understory plant species richness as the response variable. Corresponding models were fitted for the other five ecosystem services.

We assumed that the understory plant species richness followed an overdispersed Poisson distribution with mean, $\mu_{p,t}$. We modelled this mean, transformed by the ln link function, $g()$, on plot p=1, ..., $P \leq 8$, on tract t=1, ..., $T \leq 1{,}394$ as

$$g(\mu_{p,t}) = \alpha_t + \sum_{n=1}^{n} \beta_n X_{p,t} + \log(P_{p,t}) + \varepsilon_{p,t} \quad (1)$$

Where, $X_{p,t}$ is a design matrix of n plot-level explanatory variables, β_n is a vector of associated effect-size parameters, and $P_{p,t}$ is the regional species pool (see description of understory plant species richness). Explanatory variables were centred and standardized. As we included the log of the regional species pool, we in fact modelled the proportion of the regional species pool. The parameters $\varepsilon_{p,t}$ are overdispersion contributions, where $\varepsilon_{p,t} \sim N(0, \sigma)$, which means normally distributed with mean=0 and s.d.=σ. For normally distributed ecosystem services,

$\varepsilon_{p,t}$ are residual contributions. The tract-level 'intercept' parameters α_t were modelled further as

$$\alpha_t \square N\left(\mu_{\alpha_t}, \sigma_\alpha\right) \tag{2}$$

Where σ_α is the s.d., and

$$\mu_\alpha = \gamma + \sum_{m=1}^{m} \rho_m Z_t \tag{3}$$

where Zt is a design matrix of m=1, ..., M ≤ 5 tract-level explanatory variables, which were centred and standardized, m is a vector of associated effect-size parameters, and is an 'intercept' parameter.

We built the main model based on the Deviance Information Criterion (DIC52), on the posterior distribution of the effect-size parameters (n in equation (1) and m in equation (3)), and on knowledge of the biological system studied. We first assessed the predictive power of each explanatory variable based on the DIC and on the posterior distributions of n and m. Next, we fitted a multiple model containing the retained explanatory variables. Finally, we simplified this multiple model by excluding and again including earlier excluded variables in a stepwise procedure. Parameter estimates for the final main model for each ecosystem service are presented in Fig. 4. For some ecosystem services, there were a few models with similar DIC (difference in DIC<2). However, the relationships between the response variables and (i) tree species richness, and (ii) the tree species biomasses differed only marginally between these models.

For a description of parameters, symbols and prior distributions. We assumed uninformative prior distributions for all parameters. The models were fitted using the softwares OpenBUGS 2.1 (ref. 53). For the models aimed at investigating the pairwise relationships between services, we excluded all plots with missing data. Moreover, bilberry production was also arcsine transformed when used as an explanatory variable. Below follows a detailed description of our response and explanatory variables.

Ecosystem Services

Tree biomass production was estimated as the yearly change in tree biomass (kg m^{-2} $year^{-1}$), calculated over a period of 5 years, and for all tree individuals higher than 1.3 m. For plots visited in the years 1999–2002, the baseline for biomass production was thus the years 1994–1997. Biomass was calculated with biomass functions[54, 55] and was the sum of the biomass from the stem, twigs and branches, the stump and roots. For deciduous tree species, there is only a function for *Betula*spp., and this function was thus applied for all other deciduous tree species. The function for *Pinus sylvestris* was applied for *Larix decidua* and *Pinus contorta.* Even though this creates a slight tree species bias, it has minor effects on our production estimates, as we compared the biomass of the trees at two points in time. Sample size for tree production was 4,335 plots, and due to fallen trees, observed values could be negative (as was the case in 5 plots).

Soil carbon storage was measured as the amount of carbon (g m^{-2}) in the topsoil, which consisted of either purely organic horizons, for example, mor layers (63%) or peat layers (21%), or less frequently of minerogenic A-horizons (16%). This is the part of the soil most affected by the current above-ground biota. To compensate for the conceptual difference in topsoil types, mean soil carbon stocks were set equal for purely organic soils (measured in the whole organic horizon down to a maximum depth of 30 cm) and minerogenic topsoils (measured in the top 10-cm horizon). The soil fraction <2 mm was analysed. Soil sampling was carried out on around 50% of the inventory plots, totalling 1,953 plots. Data were log transformed.

Bilberry production was measured as the percentage of each plot covered by bilberry, *Vaccinium myrtillus*. The cover of *V. myrtillus* is strongly correlated to annual production, and bilberry is one of the economically most important wild berry species in Northern Europe[28]. Bilberry sampling was carried out on around 50% of the inventory plots, totalling 2,127 plots. Data were converted to proportions and arcsine transformed, after correcting for the area where berries could not grow, for example, the area of stems and boulders.

Game production potential was calculated as the percentage cover of each plot that was covered by any of the 15 plant species important for large herbivorous mammals, for example, moose (*Alces alces*) and roe deer (*Capreolus capreolus*. The cover of these field layer species was measured in the same way as the bilberry cover. Not all plants are equally preferred, however. Based on known preferences and nutritional values[57], we therefore weighted the least and most important plant species by 0.66 and 1, respectively. As the herbivores also browse on coppice, we included the cover of preferred species in this layer. Field layer and coppice were summed to yield a measure of total food for herbivores, which could amount to more than 100% because of multiple vegetation layers. Game production is the basis for both recreational hunting and meat, and was estimated to a value of US$ 460 million in Sweden as of 2006 (ref. 58). Sample size was 2,127 plots and data were log transformed after adding a constant of 0.01.

Understory plant species richness: The Swedish National Forest Inventory includes about 270 ground vegetation species, and of these, we selected those with forest as main habitat, in total 141 species (Supplementary Table S6). We modelled the proportion of the richness of these species in each plot in relation to the regional species richness of the same species, estimated as all of the 141 species occurring within a 600 × 600 km window centred on each plot. The regional species pool thus captured the change in species composition with latitude (as the longitudinal distance in the sampling region never exceeds 600 km). We found that the median number of plant species in the regional pool reached an asymptote at a distance of ~300 km from each plot, meaning that further increasing the area of the regional species pool did not, on average, incorporate more species. Sample size was 4,327 plots.

We modelled the probability of dead wood occurrence using data on presence/absence of dead wood in each plot (minimum diameter for dead wood pieces to be counted equals 40 mm). We chose this variable, as the abundance of dead wood showed a highly skewed distribution in which small fragments were common and a few plots had very high abundances. Sample size was 4,335 plots. Dead wood is the essential habitat for many organisms: more than 50% of the red-listed forest species in Sweden depend on dead wood[59].

The six ecosystem services included in our analyses are a subset of all possible services that forests provide. The results and conclusions should be viewed with this in mind.

Explanatory Variables

Temperature is defined as the total accumulated average daily temperature (in °C) over 5 °C during the vegetation period. Humidity is measured as the difference in mm between the addition of water via precipitation and the loss of water due to transpiration and evaporation during the vegetation period. For nitrogen deposition, spatially distributed N deposition data in kg N ha^{-1} $year^{-1}$ (wet and dry) were downloaded from the Swedish Meteorological and Hydrological Institute website. As temperature, humidity and nitrogen deposition are 'modelled' variables, their resolution is not at the scale of local plots, and were thus modelled at the scale of tracts. pH of the topsoil was measured in water suspension. The ratio of total carbon to total nitrogen is the ratio (C:N) in the topsoil. Soil moisture was originally measured on a categorical scale of 1 (driest) to 5 (wettest). These categories correspond to a continuous scale of % volumetric soil water contents ranging from 0.15 to 0.35 (ref. 60). Peat was modelled as peat (1) or non-peat (0) soil. The biomass of the six most abundant or regionally important ('focal') tree species in Sweden were included. These were: Norway spruce (*Picea abies*), Scots pine (*Pinus sylvestris*), birch (*Betula* spp.; *B. pendula* and *B. pubescens*), oak (*Quercus robur*), aspen (*Populus tremula*) and beech (*Fagus sylvatica*), measured in kg m^{-2}. Tree species richness was our estimate of tree species diversity. It was calculated as the observed presence of 20 species in each plot, and ranged from 1 to 10. The tree species included our six focal species together with *Acer platanoides, Acer pseudoplatanus, Alnus glutinosa, Alnus incana, Carpinus betulus, Fraxinus excelsior, Larix decidua, Pinus contorta, Prunus avium, Salix caprea, Sorbus aucuparia, Sorbus intermedia, Tilia cordata* and *Ulmus glabra*. We chose species richness as our diversity estimate because it is a simple and intuitive measure. Species richness estimates are relatively easy to obtain and relate to the ability to manage ecosystems under limited information about species' relative distributions and traits. Forest stand age was estimated in each plot as the basal area weighted mean age from at least three trees.

ACKNOWLEDGEMENTS

This work was conducted as part of the Future Forests research program in Sweden. L.G. was supported by grant 621-2009-5457 from the Swedish research council VR, and T.S. by grant 2005-933 from Formas. We thank David Wardle and Andy Hector for comments that helped improve an earlier version of the paper. Johan Mentzer helped drafting Fig. 2.

REFERENCES

1. Millenium Ecosystem Assessment. *Ecosystems and Human Well-Being: Biodiversity Synthesis*World Resources Institute. 100 (Island Press (2005).

2. Naeem S., Bunker D. E., Hector A., Loreau M., Perrings C. H. *Biodiversity, Ecosystem Functioning, & Human Wellbeing, an Ecological and Economic Perspective* Oxford University Press, (2009).

3. Cardinale B. J. *et al*. Biodiversity loss and its impact on humanity. *Nature* 486, 59–67 (2012).

4. Schmid B. *et al* in *Biodiversity, Ecosystem Functioning, & Human Wellbeing: An Ecological and Economic Perspective* eds Naeem S. Oxford University Press (2009).

5. Cardinale B. J. *et al*. The functional role of producer diversity in ecosystems. *Am. J. Bot.* 98,572–592 (2011).

6. Wardle D. A., Jonsson M. Biodiversity effects in real ecosystems—a response to Duffy. *Front. Ecol. Environ.* 8, 10–11 (2010).

7. Wardle D. A., Zackrisson O. Effects of species and functional group loss on island ecosystem properties. *Nature* 435, 806–810 (2005).

8. Reich P. B. *et al*. Impacts of biodiversity loss escalate through time as redundancy fades.*Science* 336, 589–592 (2012).

9. Naeem S., Bunker D. E., Hector A., Loreau M., Perrings C. H. in *Biodiversity, Ecosystem Functioning, & Human Wellbeing* : An Ecological and Economic Perspective (eds Naeem S Ch. 1, 3–13 Oxford University Press (2009).

10. Grace J. B. *et al*. Does species diversity limit productivity in natural grassland communities?*Ecol. Lett.* 10, 680–689 (2007).

11. Vila M. *et al*. Species richness and wood production: a positive association in Mediterranean forests. *Ecol. Lett.* 10, 241–250 (2007).

12. Paquette A., Messier C. The effect of biodiversity on tree productivity: from temperate to boreal forests. *Global Ecol. Biogeogr.* 20, 170–180 (2011).

13. Hector A., Bagchi R. Biodiversity and ecosystem multifunctionality. *Nature* 448, 188–190(2007).
14. Gamfeldt L., Hillebrand H., Jonsson P. R. Multiple functions increase the importance of biodiversity for overall ecosystem functioning. *Ecology* 89, 1223–1231 (2008).
15. Zavaleta E. S., Paṣari J. R., Hulvey K. B., Tilman G. D. Sustaining multiple ecosystem functions in grassland communities requires higher biodiversity. *Proc. Natl Acad. Sci. USA*107, 1443–1446 (2010).
16. Maestre F. T. *et al.* Plant species richness and ecosystem multifunctionality in global drylands. *Science* 335, 214–218 (2012).
17. Larigauderie A., Mooney H. A. The Intergovernmental science-policy Platform on Biodiversity and Ecosystem Services: moving a step closer to an IPCC-like mechanism for biodiversity.*Curr. Opin. Environ. Sustain.* 2, 9–14 (2010).
18. Hansen M. C., Stehman S. V., Potapov P. V. Quantification of global gross forest cover loss.*Proc. Natl Acad. Sci. USA* 107, 8650–8655 (2010).
19. Frivold L. H., Frank J. Growth of mixed birch-coniferous stands in relation to pure coniferous stands at similar sites in south-eastern Norway. *Scand. J. Forest Res.* 17, 139–149 (2002).
20. Pretzsch H., Schutze G. Transgressive overyielding in mixed compared with pure stands of Norway spruce and European beech in Central Europe: evidence on stand level and explanation on individual tree level. *Eur. J. Forest Res.* 128, 183–204 (2009).
21. Bonan G. B. Forests and climate change: forcings, feedbacks, and the climate benefits of forests. *Science* 320, 1444–1449 (2008).
22. Janisch J. E., Harmon M. E. Successional changes in live and dead wood carbon stores: implications for net ecosystem productivity. *Tree Physiol.* 22, 77–89 (2002).
23. Mace G. M., Norris K., Fitter A. H. Biodiversity and ecosystem services: a multilayered relationship. *Trends Ecol. Evol.* 27, 19–26 (2012).
24. Cotta H.v. *Anweisung zum Waldbau* Arnoldische Buchhandlung (1828).

25. Pretzsch H. in *Forest Diversity and Function* Vol. Ecological Studies eds Scherer-Lorenzen M., Körner C., Schulze E. D. 41–64Springer-Verlag (2005).

26. Hooper D. U., Vitousek P. M. Effects of plant composition and diversity on nutrient cycling.*Ecol. Monogr.* 68, 121–149 (1998).

27. Duffy J. E., Richardson J. P., Canuel E. A. Grazer diversity effects on ecosystem functioning in seagrass beds. *Ecol. Lett.* 6, 637–645 (2003).

28. Miina J., Hotanen J. P., Salo K. Modelling the abundance and temporal variation in the production of bilberry (Vaccinium myrtillus L.) in Finnish mineral soil forests. *Silva. Fenn.* 43,577–593 (2009).

29. Food and Agricultural Organization of the United Nations. in *Global Forest Resources Assessment 2010: Main Report* FAO Forestry Paper Rome (2010).

30. Swedish Forest Agency. *Statistical Yearbook of Forestry 2012: Official Statistics of Sweden*Jönköping (2012).

31. Anderson R. C., Loucks O. L., Swain A. M. Herbaceous response to canopy cover light intensity and throughfall precipitation in coniferous forests. *Ecology* 50, 255–263 (1969).

32. Nelson E. *et al.* Modeling multiple ecosystem services, biodiversity conservation, commodity production, and tradeoffs at landscape scales. *Front. Ecol. Environ.* 7, 4–11 (2009).

33. Raudsepp-Hearne C., Peterson G. D., Bennett E. M. Ecosystem service bundles for analyzing tradeoffs in diverse landscapes. *Proc. Natl Acad. Sci. USA* 107, 5242–5247 (2010).

34. Schnitzer S. A. *et al.* Soil microbes drive the classic plant diversity-productivity pattern.*Ecology* 92, 296–303 (2010).

35. Godbold J. A., Solan M. Relative importance of biodiversity and the abiotic environment in mediating an ecosystem process. *Mar. Ecol. Prog. Ser.* 396, 273–282 (2009).

36. Dise N. B., Rothwell J. J., Gauci V., van der Salm C., de Vries W. Predicting dissolved inorganic nitrogen leaching in European forests using two independent databases. *Sci. Total Environ.* 407, 1798–1808 (2009).

37. Bulling M. T. *et al.* Marine biodiversity-ecosystem functions under uncertain environmental futures. *Phil. Trans. R. Soc. B* 365,

2107–2116 (2010).

38. Hickler T. *et al.* Projecting the future distribution of European potential natural vegetation zones with a generalized, tree species-based dynamic vegetation model. *Global Ecol. Biogeogr.* 21, 50–63 (2012).

39. Scherer-Lorenzen M., Körner C., Schulze E. D. *Forest Diversity and Function* Vol. 176, (Springer-Verlag (2005).

40. Morin X., Fahse L., Scherer-Lorenzen M., Bugmann H. Tree species richness promotes productivity in temperate forests through strong complementarity between species. *Ecol. Lett.*14, 1211–1219 (2011).

41. Caspersen J. P., Pacala S. W. Successional diversity and forest ecosystem function. *Ecol. Res.* 16, 895–903 (2001).

42. Worm B. *et al.* Impacts of biodiversity loss on ocean ecosystem services. *Science* 314,787–790 (2006).

43. Danovaro R. *et al.* Exponential decline of deep-sea ecosystem functioning linked to benthic biodiversity loss. *Curr. Biol.* 18, 1–8 (2008).

44. Mora C. *et al.* Global human footprint on the linkage between biodiversity and ecosystem functioning in reef fishes. *PLoS Biol.* 9, e1000606 (2011).

45. Ptacnik R. *et al.* Diversity predicts stability and resource use efficiency in natural phytoplankton communities. *Proc. Natl Acad. Sci. USA* 105, 5134–5138 (2008).

46. Isbell F. *et al.* High plant diversity is needed to maintain ecosystem services. *Nature* 477,199–202 (2011).

47. Foley J. A. *et al.* Solutions for a cultivated planet. *Nature* 478, 337–342 (2011).

48. Hooper D. U. *et al.* A global synthesis reveals biodiversity loss as a major driver of ecosystem change. *Nature* 486, 105–108 (2012).

49. Shepherd G. *The Ecosystem Approach: Learning from Experience* IUCN (2008).

50. Axelsson A.-L. *Et al* in *National Forest Inventories—Pathways for Common Reporting* eds Tomppo E., Gschwantner Th., Lawrence M., McRoberts R. E. 541–553Springer (2010).

51. Gelman A., Carlin J. B., Stern H. S., Rubin D. B. *Bayesian Data Analysis* 2nd edn CRC Press (2004).
52. Spiegelhalter D. J., Best N. G., Carlin B. R., van der Linde A. Bayesian measures of model complexity and fit. *J. R. Stat. Soc. Ser. B Stat. Methodol.* 64, 583–616 (2002).
53. Thomas A., O'Hara R. B., Ligges U., Sturtz S. Making BUGS open. *R News* 6, 12–17 (2006).
54. Marklund L. G. *Biomass Functions for Pine, Spruce and Birch in Sweden* Department of Forest Survey, SLU (1988).
55. Petersson H., Ståhl G. *Functions for Below-Ground Biomass of Pinus Sylvestris, Picea Abies, Betula Pendula and Betula Pubescens in Sweden* Department of Forest Resource Management and Geomatics, SLU (2006).
56. Cederlund G., Ljungqvist H., Markgren G., Stålfelt F. Foods of moose and roe deer at Grimsö in central Sweden—results of rumen content analysis. *Swed. Wildlife Res.* 11, 167–247(1980).
57. Hanley T. A. A nutritional view of understanding and complexity in the problem of diet selection by deer (Cervidae). *Oikos* 79, 209–218 (1997).
58. Boman M., Mattsson L., Ericsson G. R., Kriström B. Moose hunting values in Sweden now and two decades ago: the Swedish hunters revisited. *Environ. Resource Econ.* 50, 515–530(2011).
59. National Atlas of Sweden. The Royal Swedish Academy of Agriculture and Forestry.*Agriculture and Forestry in Sweden since 1900 – a Cartographic Description* 232 (Stockholm (2011).
60. Warfvinge P., Sverdrup H. *Critical Loads of Acidity to Swedish Forest Soils. Methods, Data and Results* 104 (Lund (1995).

Chapter 3

An Overview of Indian Forestry Sector with REDD+ Approach

Vandana Sharma and Smita Chaudhry

Institute of Environmental Studies, Kurukshetra University, Kurukshetra 136119, Haryana, India

ABSTRACT

Forest ecosystems cover large parts of the terrestrial land surface and are major components of the terrestrial carbon (C) cycle. The primary objective of REDD+ is to minimize the carbon emissions from deforestation in developing countries and enhance their carbon storage capacities through sustainable management programme. The recognition of REDD+ throughout the international community, its support by donors and promotion in the perspectives of the UNFCCC negotiations are mainly due to vital functions of forests in regulating the world's climate. This paper gives an overview of REDD+ approach and its methodological guidance in context of Indian forestry sector. The strengthening of governance arrangements and institutions in India needs to integrate learning through piloting, adaptive management, and knowledge transfer. A phased approach for India for REDD+ implementation having safeguards for local communities and biodiversity along with a system of their reporting and capacity building has to be developed. Successfully designed REDD+ implementation in India entirely depends on a rigid, scalable, and reliable finance mechanism, technological assistance, and effective forest-related legislation along with transparent and equitable political momentum which has support of core stakeholder groups.

INTRODUCTION

Forests, like other ecosystems, are affected by climate change. Forests also influence climate, absorbing CO_2 from the atmosphere and storing carbon in wood, leaves, litter, roots, and soil. The carbon is released back into the atmosphere when forests are cleared or burnt. By acting as sinks, forests are considered to moderate global climate change [1]. Climate change is one of the most significant global challenges of our time, and addressing it requires the urgent formulation of comprehensive and effective policy responses [2]. Natural forests are more resilient to climate change and disturbances than plantations because of their genetic, taxonomic, and functional biodiversity. This resilience includes regeneration after fire, resistance to and recovery from pests and diseases, and adaptation to changes in radiation, temperature, and water availability (including those resulting from global climate change). While the genetic and taxonomic composition of forest ecosystems changes over time, natural forests will continue to take up and store carbon as long as there is adequate water and solar radiation for photosynthesis [3]. Forests play a major role in the global carbon (C) cycle because they store 80% of the global aboveground C of the vegetation and about 40% of the soil C and interact with atmospheric processes through the absorption and respiration of CO_2 [4–8]. Human activities such as fuel consumption and land-use change are the main causes of an increase in the atmospheric carbon dioxide concentration, which is generally recognized as a factor of climate change and global warming [9]. The Stern Review [10] highlighted that forest conservation, afforestation, reforestation, and sustainable forest management can provide up to 25% of emission reductions needed to effectively combat climate change and that curbing deforestation has the potential to offer significant emission reductions fairly quickly in a highly cost-effective manner. Reducing emissions from deforestation and forest degradation (REDD+) is a mechanism for providing financial rewards to countries that reduce carbon emissions caused by the loss and degradation of their forests. In concept, REDD resembles other Payment for Environmental Services (PES) programs; however, REDD emphasizes a reduction in deforestation and degradation rates from expected levels, also known as avoided deforestation and degradation [11].

India has a total land area of 329 million hectares [12] of which around 23.4%, that is, 76.87 million hectares (Mha), is classified as the forestland with tree cover. More than 40% of the country's forests are degraded and understocked [13, 14]; subsequently there is a large potential of REDD+ activities in the country. Hence an overview of REDD+ approach along with its methodological guidance is discussed in this article in context of Indian forestry sector.

RED TO REDD+: EVOLUTION SINCE ITS CONCEPTUALIZATION

Fuelled by the continuing destruction of forests in developing countries, the aspired development and implementation of a REDD+ mechanism under the United Nations Framework Convention on Climate Change (UNFCCC) evolved into one of the major issues in the negotiations on a post-Kyoto agreement [15]. The concept of REDD was first introduced in its preliminary form at the climate change negotiations during the third meeting of the Conference of Parties to the UNFCCC (COP-3) in 1997 to enlist carbon services of forests under the Clean Development Mechanism (CDM) and account for emissions and removals from land use, land use change, and forestry (LULUCF) activities [16]. Reducing emissions from deforestation in developing countries was being discussed in the side events since COP-9 of the UNFCCC in 2003 under the names of "Avoided Deforestation," "Compensated reduction," and so forth [17]. When the Kyoto Protocol came into force in 2005, it focused on reductions in emissions from technological projects. This expression was used for the first time in its shortened form RED (reducing emissions from deforestation) during the 11th UN Conference of Parties (COP-11) in Montreal (2005) by the Coalition for Rainforest Nations led by Papua New Guinea—noting that deforestation was estimated to account for 12–15% of the overall greenhouse gas (GHG) emissions [18]. Well received at COP-11, the concept was further elaborated, expanded, and officially adopted during COP-13 in Bali, Indonesia, in 2007 in the form of REDD. The addition of "degradation" to this acronym was due to the observation that forest degradation in some developing countries was as threatening as deforestation (if not more) to the forest ecosystems. Following the debates during the 14th COP in Poznan, Poland, in 2008, it was decided that REDD should evolve to REDD+ to encompass all the initiatives that can increase the carbon absorption potential of forests.

The insertion of "+" on the acronym REDD is aimed at broadening its scope to include all operations associated with preservation, restoration, and sustainable management of forest ecosystems. The official definition of REDD+ as set by UNFCCC is "reducing emissions from deforestation and forest degradation in developing countries, and the role of conservation, sustainable management of forests, and enhancement of forest carbon stocks in developing countries [19]. During COP-16 of UNFCCC in 2010 in Cancun, governments agreed to boost action to curb emissions from deforestation and forest degradation in developing countries with technological and financial support [17]. REDD+ attaches financial value to the carbon stocks stored in the forests and other incentives for developing countries by the developed countries to reduce emissions and invest in low carbon paths to sustainable development. It goes beyond deforestation and forest degradation and includes the role of conservation, sustainable management of forests, and enhancement of carbon stocks. COP-17 in November, 2011 at Durban, South Africa, produced a landmark decision to extend the Kyoto Protocol into the second commitment period. Pursuant to Cancun decision, important decisions on REDD+ were made, namely, decisions on systems for providing information on safeguards, modalities for forests reference (emission) levels, and REDD+ financing [20]. The primary goal of REDD+ is reduction of greenhouse gas emissions, consistent with the goal of the UNFCCC to achieve "stabilization of greenhouse gas concentrations in the atmosphere at a level that would prevent dangerous anthropogenic interference with the climate system." REDD+ is expected to bring much more than emission reductions, with a properly designed mechanism contributing to multiple benefits. Depending on the location and type of REDD+ activity, these benefits potentially include poverty alleviation, indigenous people's rights, improved community livelihoods, technology transfer, sustainable use of forests resources, and biodiversity conservation [21].

REDD+ is a complex instrument of governance. It is treated by many stakeholders as a payments for ecosystems services (PES) scheme [22, 23] which seeks to translate the process of carbon sequestration through arboreal photosynthesis into real financial incentives for ecosystems managers and in so doing to promote the conservation and sustainable management of natural habitats [24, 25]. The implementation of REDD-plus requires forest-governance reforms through inclusive processes that build on existing forest-governance systems [26].

India has played an important role in REDD+ negotiations and has been instrumental in shaping the REDD+ mechanism by emphasising the role of conservation and sustainable forest management in mitigating carbon emissions [16].

GREEN FACTS OF INDIA: A MEGADIVERSE COUNTRY

The forests of India have long been an important part of its culture and a defining feature of its landscape [27]. India with a wide range of climate, geography, and culture is unique among biodiversity-rich nations and is known for its diverse forest ecosystems and megabiodiversity. It ranks as the 10th most forested nation in the world [28], with 23.4% (76.87 Mha) of its geographical area under forest and tree cover [1, 29]. Out of 34 global biodiversity "hot spots," four are located in India that is, Eastern Himalayas, North-east, Sundarbans, and Western Ghats [30]. India is one of the 17 megadiverse countries (MoEF). Fifteen biodiversity-rich areas of country covering an area of approximately 74000 km^2 have been designated as biosphere reserve and four of them, namely Nilgiri, Nanda Devi, Sundarbans, and Gulf of Mannar, have been recognised by UNESCO underworld network of biospheres [31]. With only 2.4% of the land area, India accounts for 7 to 8 percent of the recorded species of the world [32]. This biodiversity is of immense economic, ecological, social, and cultural value. Approximately 275 million people in India (27% of the total population) are known to live in the forest fringes and earn bulk of their livelihood from forests [33–36]. FSI defines forests as "all the lands, more than one hectare in area, with a tree canopy density of more than 10%." Champion and Seth [37] classified India's forests into four major ecosystems groups, namely, tropical, subtropical, temperate, and alpine. These major groups are further divided into 16 types. Of the 16 forests types, tropical dry deciduous forests form the major percentage that is 38% of the forest cover in India.

Key Challenges to Indian Forestry Sector

India's economic growth in the last decade has raised several concerns in terms of its present and future resource demands for material and energy [38]. With 18% of global livestock population over 2.4% of world's geographic area and 17% of world's population rigorous biotic pressure being faced by Indian forests as only due to these forests, nearly 30% of fodder needs of the cattle population and 40% of domestic fuel wood needs of the people are being catered. The difference in demand and supply is also broadening for fuel wood, timber, and fodder. In Eastern and Northeastern India, forest degradation is mainly a result of shifting cultivation practices over an area about 1.2 Mha. The National Forest Commission had done a splendid task to conquer the issues forests are facing by recommending allocation of a minimum 2.5% of national budget to the forestry sector. Apprehensions over inadequate role of elected Panchayati Raj institutes vis-a-vis Joint Forest Management Committees (JFMCs) in forest management, limited participation of non-profit-making voluntary sector, control over minor forest products, and implementation of Forest Right Act and PESA are some concerns frequently voiced with little recognition for the incredible efforts made to maintain forest cover in the current conflicting scenarios [31]. As per the wood budget for the year 1996, 86 million tonnes of fuel wood is being unsustainably removed from forests [39]. India's livestock population of 467 million grazes on 11 Mha of pastures. This implies that an average of 42 animals graze on a hectare of land compared to a threshold level of 5 [40]. In absence of adequate grazing land, nearly a third of the fodder requirement is met from the forest resources in the form of grazing and cut fodder for stall feeding. An estimated 100 million cow units graze in forests annually whereas the sustainable level is only 31 million [41]. Additionally, graziers collect an estimated 175–200 million tonnes of green fodder annually [42]. Grazing has been reported in 67% of the national parks and 83% of the wildlife sanctuaries surveyed [40, 43].

Deforestation

Around 3000 B.C, nearly 80% of India was forested [44, 45]. Deforestation has occurred in the tropics throughout history [46, 47]. Growing population, widespread poverty, and limited employment opportunities in agricultural and industrial sector have resulted in heavy pressure on forests, primarily due to unsustainable extraction of fuel wood and over grazing resulting in forest degradation [48]. Forest vegetation sequesters carbon while at the same time deforestation and degradation of standing forests leads to release of stored carbon [49]. The total forest cover has been declining globally. It stood at 3.95 Mha or about 30% of the global land area in 2006. The gross deforestation rate is estimated to have declined between 2000 and 2005 compared to the decade of the 1990s, but it still amounted to a gross and net loss of 12.9 and 7.3 Mha/year, respectively [50]. Deforestation refers to conversion of area having forest cover to other uses, for example, croplands, pastures, or urban land. Degradation [51], on the other hand, refers to reduction in productivity and/or diversity of a forest due to unsustainable harvesting (removals exceeding replacements and changes in species composition), fire (except for fire-dependent forest systems), pests and diseases, removal of nutrients, and pollution or climate change (e.g., changes in productivity, total organic matter, and forest composition). In India, the forest cover is now relatively stable and, therefore, deforestation is not currently a significant issue in formulation of carbon sequestration policies [40].

Stabilization of Carbon Stocks in Indian Forests

India recognized the enormous importance of the forest resources and land use, land-use change, and forestry (LULUCF) activities in contributing towards GHG emissions. Strong policy skeleton in India made the conservation of forests more focused [17]. The forestry sector recognizes its increasing role to provide sustained benefits to the people and strives to attain it by integrating new frontiers of knowledge and science in planning, management, research and capacity building with forest management [31]. A major part of forest land at the time of independence came under reserved forests [21]. During 1952–1976, forests were recognized only for the commercial interest, (Table 1), that is, for eradication of valuable timber for fast growth of industrial sector, nation's development and other purposes [36, 52].

Table 1: Area under different legal categories of forest in India during 1946-47 [36, 52]

Legal status	**Area (million ha)**	**% of total area**
Reserve forest	25.32	96.79
Protected forest and unclassified forest	0.84	3.21
Total	26.16	100

Table 2: presents the change in forest cover over the years 1987 to 2011 indicating that the forest cover has increased marginally from 64.08 million ha in 1987 to 69.2 million ha in 2011

Year	Recorded forest area (in Mha)					Forest cover (in Mha)				
	Reserved forests	Protected forests	Unclassed forests	Total forests area	% of geographical area	Dense forests	Open forests	Mangroves	Total	% of geographical area
1987	40.18	21.73	13.27	75.18	22.8	36.14	27.66	0.4	64.08	19.49
1989	40.18	21.73	13.27	75.13	22.8	37.84	25.74	0.42	63.88	19.43
1991	41.49	23.30	12.20	77.008	23.4	38.50	24.99	0.42	63.93	19.45
1993	41.49	23.30	12.20	77.008	23.4	38.55	25.02	0.42	63.93	19.45
1995	41.65	22.33	12.53	76.52	23.38	38.57	24.93	0.45	63.88	19.43
1997	41.65	22.33	12.53	76.52	23.38	36.72	26.13	0.48	63.33	19.27
1999	41.65	22.33	12.53	76.52	23.38	37.73	25.50	0.48	63.72	19.39
2001	42.33	21.72	12.78	76.84	23.38	41.68	25.87	0.45	67.55	20.55
2003	39.99	23.84	13.63	77.47	23.57	39.05	28.77	0.44	67.83	20.64
2009	43.05	20.62	13.27	76.95	23.41	40.25	28.84	0.46	69.09	21.02
2011	42.25	21.39	13.30	76.95	23.81	40.42	28.78	0.46	69.20	21.05

India is one of the few countries where deforestation rates have been reduced and regulated and forest cover has nearly stabilised, unlike most other developing countries [28]. Thus it is important to understand the likely factors contributing to the observed stabilization of forest carbon stocks in India. The factors include legislations, forest conservation and afforestation programmes, and community awareness and participation [64]. Realizing the need for conservation and regeneration, several programmes have been implemented at the government and nongovernment levels. The forest policy in India was first introduced in colonial period, and later on changes were made in postcolonial period.

The few important acts, instruments, and rules governing the protection and conservation of forests include The Indian Forest Act, 1927; National Forest Policy, 1952; The Indian Wildlife (Protection) Act, 1972, amended in 1993; Forest (Conservation) Act, 1980, amended in 1988; Forest (Conservation) Rules, 1981, amended in 1992; The National Forest Policy, 1988; Joint Forest Management (JFM), 1990; Biological Diversity Act, 2002; Forest (Conservation) Rules, 2003; Biological Diversity Rules, 2004; and National Environment Policy, 2006. Of these, major contributory acts which majorly caused the stabilization of Indian forest cover are the following.

1. Forest (Conservation) Act, 1980, amended 1988: this act is one of the most effective pieces of legislation contributing to reduction in deforestation. This was enacted to reduce the discriminate diversion of forest land for nonforestry purposes and to help regulate and control the land-use changes in the existing forest area. With this act, the deforestation and conversion of forest lands to nonforest use were effectively checked [64]. The rate of conversion of forests to nonforestry uses has declined drastically to around 15,500 ha per annum since 1980 compared to 1,50,000 ha per annum prior to 1980 [42]. This act stipulates prior central government approval before any forest land is sought to be diverted for nonforestry purpose [40].
2. The National Forest Policy, 1988: India's National Forest Policy was formulated four years before the Earth Summit and embodies all elements—social, environment, and economic—of sustainable forest management. Its aim was to ensure environmental stability and maintenance of ecological balance, including atmospheric

equilibrium, which is vital for sustenance of all life forms—humans, animals, and plants [16].

3. Joint Forest Management (JFM), 1990: The Forest Policy 1988 set the stage for participatory forest management in India. The JFM programme documented the rights of the protecting communities over forests lands. The local communities and the forest department together plan and implement forest recognition programmes and the communities are rewarded for their efforts in protection and management. The total area covered under JFM programme is over 15 Mha. This has allowed protection of forests plantation, potentially contributing to conservation of existing forests and carbon stocks [64].

Forest Degradation

Forest degradation is a complex process, and its drivers may be completely different from those for deforestation, thus presenting greater challenges in assessment or monitoring as compared to deforestation [65]. Widespread forest degradation in the developing countries remains poorly understood or quantified [66]. Change in forest composition because of selective overexploitation, loss of natural regeneration, low growing stock, and low productivity are important parameters resulting in low carbon content in the existing forests [40]. Shifting cultivation, fires, and over-grazing have resulted in the elimination of susceptible species and in making selected tolerant species more abundant [51]. The dynamics and causes of deforestation and forest degradation are multifaceted and complex and vary from place to place.

Drivers of deforestation and forest degradation in Indian perspective fall into two categories: first, who those are planned and projected in accordance with policies, legal framework, management plans, and so forth, and second, those who are spontaneous, beyond government and management control, and usually not accounted for. Planned and unplanned withdrawals from forests require proper understanding and management tools including transparent governance, effective enforcement, and appropriate mitigation actions [67]. The distinction between direct and underlying causes and between human- and naturally induced change is often not as clear as it might appear. In reality, there are long, complex chains of causation that can bring about deforestation or the degradation of forests [68].

If appropriate policy instruments and management options including effective legal framework and site specific mitigation measures are introduced then the impacts of planned or controlled drivers could be minimized. Challenge lies in addressing and managing unplanned drivers and activities which are mainly a direct outcome of local people's dependence on the adjoining forest areas to meet their livelihood needs like necessity of fuel wood, fodder, grazing, food supplements, and to a very small extent an illegal mining within forest and so forth [67]. Forest productivity is the net annual increment per unit forest area. The productivity of Indian forests is low [40]. The growing stock of forests area is estimated to be around 58.96 $m^3 ha^{-1}$ in ISFR for the year 2009 to 2011, which is far lower when compared to the global average of 130.7 $m^3 ha^{-1}$ and South and Southeast Asian average of 98.6 $m^3 ha^{-1}$ for the corresponding period [69, 70]. The National Forest Commission report 2006 indicated that around 41 per cent of total forest in the country is already degraded, 70 percent of the forests have no natural regeneration, and 55 per cent of the forests are prone to fire [71]. As per the study of Nayak et al. [70] the factors affecting forest degradation in India mainly include critical livelihood, demand and supply gap of forests products, forests fires, overgrazing, illegal felling, and diversion of forest land for nonforest uses due to competing land-use demand for development and so forth [13, 61, 63,71–73].

Forest Degradation and Deforestation: Addressal Mechanism

Reserving forests implies foregoing the benefits that would have been generated by exploiting the resources or from adopting alternative land-use practices. Further, the on-site benefits of forests are lower than the potential benefits of alternative land uses [16]. The strategies and approaches to evolve a mechanism to tackle forest degradation and deforestation will be critical to attain the maximum benefits of incentives for the stored carbon stocks. Creation of appropriate awareness amongst stakeholders in deciding the level of their participation and commitment including that of the local community for protection of existing forests might be the leading action for checking deforestation and forest degradation. The stakeholders need to be effectively and appropriately informed, enthused, and empowered to take apt decision for conservation of forests. This should be supported by the government programmes and policies, which should advocate and provide the relevant alternative resources to the community so that the dependence of local community on forest resources could be reduced. As per Indian submission to SUBSTA, UNFCCC, these programmes could include

1. alternative cheap cooking fuel supply,
2. promoting non-conventional energy sources,
3. low-cost permanent housing facilities,
4. improving agriculture and livestock productivity,
5. free education for children,
6. better infrastructural facilities including health,
7. effective use of modern communication audiovideo tools for creating awareness among community.

Striking a balance between the need to enhance food production for growing population and to cut short deforestation requires increase in agricultural production without further deforestation. This can be addressed through better land planning and extensive investment in technology to increase yields of existing farmlands [67].

SAFEGUARDS FOR REDD+ IMPLEMENTATION

Since 2009, the UNFCCC negotiations have increasingly taken up the concerns regarding potential negative effects of REDD+ on the biodiversity of forest ecosystems, safeguards and benefits being key words in this matter [15]. The implementation of REDD+ actions can pose a number of risks or negative impacts including conversion of natural forests to plantations and other land uses of low biodiversity, loss of traditional territories, erosion of rights with exclusion from land, and loss of traditional livelihoods [74]. At COP-17 in Durban (Decision 12/CP.17 Para 5) it was agreed that a summary of information on how safeguards are being addressed and respected should be provided periodically in national communication channels agreed on by the COP [16].

The new term safeguards was introduced during AWG-LCA intersessions (Bonn, August 2009): "...[safeguards to protect biological diversity in host countries, including safeguards against conversion of natural forests to forest plantations, should be established]." The underlying principle behind this was to deal with the dilemma of not having a mandate to explicitly include biodiversity and the simultaneous need to ensure that REDD would not create incentives that could offset the biodiversity objectives of the CBD. In later AWG-LCA negotiations cobenefits were replaced by safeguards (nonpaper 18 and 39, FCCC/AWGLCA/2009/L.7/Add.6). At COP15 the text on this matter reads and "... the following safeguards should be [promoted] [and] [supported] (e) Actions that are consistent with the conservation of natural forests and biological diversity, ensuring that actions [...] are not used for the conversion of natural forests, but are instead used to incentivize the protection and conservation of natural forests and their ecosystem services, and to enhance other social and environmental benefits" [15].

In many REDD+ countries, discussions on safeguards are in their infancy and represent only a minor component of the overall REDD+ policy dialogue [75]. Safeguards can be most effectively addressed if explicit consideration is given to biodiversity concerns during all of the planning and design, implementation, and assessment stages of the REDD+ process [76]. The REDD+ framework that is part of Cancun Agreements includes a number of safeguard provisions that are to be addressed and respected throughout the implementation of REDD+ activities [17]. The UNFCCC, in the Cancun Agreement, articulated seven social and environmental safeguards for REDD+ [77]. These are

1. actions that complement or are consistent with the objectives of national forest programmes and relevant international conventions and agreements;
2. transparent and effective national forest governance structures, taking into account national legislation and sovereignty;
3. respect for the knowledge and rights of indigenous peoples and members of local communities, by taking into account relevant international obligations, national circumstances and laws, and noting that the United Nations General Assembly has adopted the UN Declaration on the Rights of Indigenous Peoples;
4. the full and effective participation of relevant stakeholders, in particular indigenous peoples and local communities;

5. actions that are consistent with the conservation of natural forests and biological diversity, ensuring that actions referred to in paragraph 70 of this decision are not used for the conversion of natural forests but are instead used to incentivize the protection and conservation of natural forests and their ecosystem services and to enhance other social and environmental benefits;
6. actions to address the risk of reversals;
7. actions to reduce the displacement of emissions.

The Subsidiary Body for Scientific and Technical Advice (SBSTA) has been tasked with providing guidance on systems for information on how REDD safeguards are being addressed and respected throughout the implementation of REDD+ [78]. Implementation of safeguards should be country based and not imposed externally. Safeguards need to be flexible and reflect national status and not formulated as additionality [17]. Free, prior, and informed consent (FPIC) could be an important element of REDD+ safeguard programs, ensuring that activities are implemented in a way that fully respects the rights of affected communities [21,79].

In order to effectively monitor safeguards in India, it will be necessary to evolve a clearly defined set of indicators and criteria for parameters such as forest governance structures, respect for rights of indigenous peoples, and full and effective participation of relevant stakeholders, along with a system to monitor these [20,29]. India has well-established forest governance system, responsible for forest management as per the Indian government policy, legal institutions, and regulatory framework. The governance system impounds the dynamism of forest behaviour. The management of forests gets inputs from the various local, regional, and national level bodies, constituting with members from the local villages. In India, there are safeguards already in place to protect the customary rights and traditions of tribes, forest dwellers, and other local communities. Policy and legal instruments exist in the form of joint management programmes, the Forest Rights Act, and the Biological Diversity Act, whose provisions ensure the rights of local communities and enable them to be key players in the local level governance of the natural resources [16].

DEVELOPING MRV SYSTEM FOR REDD+

Transparent, accountable, and sustainable mo nitoring, reporting, and verification (MRV) systems are essential for any REDD+ framework. With the prospect of a global agreement on forest preservation on the horizon, establishing functional MRV systems is one of the major goals of the so-called REDD Readiness [80]. MRV relates to both actions on the ground (i.e., that change forest carbon stocks) and REDD+ transactions (i.e., compensation and financial transactions or transfers) [81]. A robust monitoring mechanism is essential for successful implementation. As part of the Cancun Agreement, countries are supposed to develop a robust and transparent national forest-monitoring system with the capacity to consistently and accurately monitor changes in forest cover and carbon stocks over time. There is a general consensus that this forest monitoring system would be a combination of remote sensing and ground-based systems [20, 65]. Any system that intends to compensate countries for avoided deforestation and degradation requires a reference level for recent forest carbon stocks against which future performance can be measured. The choice of reference level will determine not only the effectiveness of REDD+ in terms of climate impact, but also the potential financial benefits to recipients of REDD+ funding and hence incentives for countries to participate in the scheme [82, 83].

India is among the few countries to regularly use satellite-based remote sensing technology since the 1980s in detecting and assessing forest cover changes [63]. Under the second National Communication to UNFCCC, forest biomass carbon and soil organic carbon were estimated in India, and the procedure which was followed for the programme has the potential of being developed and adopted as a REDD+ methodology for assessing changes in forest carbon stocks over a stipulated period [17]. India is of the view that the reference level (RL)/reference emission level (REL) needs to be fixed in an open and transparent manner following the procedure decided by the parties for the purpose [16]. India will have to develop a robust system for monitoring of carbon stocks in forests through a network of permanent monitoring plots to provide adequately accurate subnational estimates of carbon stock changes [20].

The Government of India formally adopted Community-based Forestry Management/Joint Forest Management resolution on July 1, 1990, in pursuance of its National Forest Policy, 1988. It laid down broad guidelines for an institutional arrangement involving the local people to jointly protect and manage the forest resources in return for benefits from it [40]. Joint forest management (JFM) has facilitated protection and regeneration of existing forests and rise of forest plantations, which will contribute to conservation of existing forests as also the carbon stocks. This approach matches well with the objectives of REDD+ programmes being implemented in many REDD+ countries [17].

FINANCING OF REDD+ MECHANISM

Essentially, REDD+ is an investment focusing on retaining or enhancing natural capital and provides an opportunity to enable countries to move towards realizing green development [84]. Multilateral, bilateral, and private funding mechanisms are now supporting different REDD+ activities at various levels. Multilateral mechanisms like UN-REDD programme, World Banks' Forest Caron Partnership Facility (FCPF), and Forest Investment Programme (FIP) are supporting capacity building activities in many developing countries for the effective implementation of REDD+ [85]. REDD+ investments are focused on maintaining or enhancing natural capital, either through investments in forests or through slowing, halting, or reversing drivers of deforestation and forest degradation [84]. REDD+ finance mechanism must be effective: they must contribute to tangible and independent third-party-verifiable stabilizations of atmospheric concentrations of greenhouse gases; efficient: they must result in "value for money" and allow both private-sector and public-sector institutions to participate on fair terms; and equitable: at a minimum, they must avoid exposing to greater risk the poor and most marginalized rural communities whose livelihoods depend on forests, avert the distortion of forest products markets, and allow broad participation on equitable terms at the national and international levels [26].

India supports a mix of market and global funds to finance REDD+ activities. Central funding should compensate for maintenance of forest carbon stocks whereas money for compensating change in carbon stocks (due to decrease in deforestation and degradation or increase in forest cover) could be generated by selling carbon credits in the international markets [73, 85]. Separate financial approaches need to be adopted for providing positive incentives for the two types of carbon stocks under REDD+ regime, that is, for (a) change in carbon stocks (with subcategories for incremental carbon stocks and reduced deforestation) and (b) baseline carbon stocks [17]. The market-based approaches that would be developed for incentivizing removals and emission reductions to be separate from the CDM market and conservation of forest carbon stocks could be incentivized through non-market-based mechanisms [16]. Any REDD finance mechanism—market-based or fund based—needs to properly address the concepts of carbon stock and flow, uncertainty, and discount rates [26].

Green India Mission

In the context of the overall objective of the National Action Plan on Climate Change (NAPCC), the environmental service which is of utmost importance, is the carbon sequestration potential of the forests. NAPCC correctly recognizes increase and improvement in forest and tree cover as a potential mitigation option. Indian initiative to have a "national mission for a green India" as one of the eight national missions is a very pragmatic step that fits most appropriately into the country's concern on climate change, as also into its overall developmental planning [86]. The government has put in place Green India Mission with a budget of Rs 46,000 crores (approximately USD 10 billion) over a period of 10 years. The mission will help in improving ecosystem services in 10 million ha of land and increase flow of forest-based livelihood services and income of about 3 million forest dependent households [27].

The GIM acknowledges the influence that the forestry sector has on environmental amelioration and inclusive development [87]. The Green India Mission puts the "greening" in the context of climate adaptation and mitigation, aiming to enhance ecosystem services like hydrological services and biodiversity, including carbon sequestration and storage (in forests and other ecosystems), in addition to provisioning services like fuel, fodder, timber, and NTFPs, while also addressing the livelihood issues of people living in and around forests [88]. In addition to other objectives, the mission will also seek positive incentives from

REDD+ mechanism. The experience of implementing concept of JFM can successfully be replicated in protecting the rights of the tribal and other forest-dwelling communities while involving them in the proposed REDD+ mechanism. The mission is the practical demonstration of India's support for policy of conservation, sustainable management of forests and increase in forest cover as a means of reducing emissions from deforestation. However, achievement of incremental targets laid in the mission would not be possible fully unless the mission receives the supplemental financial support from the future REDD+ mechanism. Similarly, there is likelihood of financial support for maintenance of baseline stocks through REDD+ mechanism [86].

The Forest Carbon Partnership (FCPF) aims to prepare forest countries for REDD+ implementation—known as REDD+ Readiness [17]. The FCPF assists developing countries in their efforts to reduce emissions from deforestation and forest degradation and foster conservation, sustainable management of forests, and enhancement of forest carbon stocks (all activities commonly referred to as "REDD+") by providing value to standing forests. The Interim REDD+ Partnership created in Paris during the May 2010 conference on forests and climate change was made up of an initial group of six developed nations who have pledged to provide US$4.5 billion to assist developing countries jumpstart REDD+ activities [19]. The Eliasch Review [89] estimated that the one-time cost requirements for 40 countries to reform policy and build capacity for REDD totalled US$4 billion.

Whether REDD is financed through a voluntary fund, a market mechanism, a hybrid mechanism (e.g., revenues from permit auctions in a cap and trade system), or a combination thereof, it must generate finance at the appropriate scale and to the appropriate stakeholders and sustain it over time [90].

REDD+ STRATEGY OF INDIA

The REDD+ text in the Cancun Agreements has come out from the text that had been discussed for many years in the UNFCCC negotiations. The text sets the stage for a nationally driven phased approach to a REDD+ mechanism. The international framework involves a three-phase approach for an REDD+ mechanism for developing countries like India: (1) development of national strategies or action plans and capacity building; (2) implementation of national strategies or action plans that could involve REDD+ pilot projects; and (3) mobilization of funds from developed countries, with financing mechanisms [21].

The REDD+ strategy plan for India containing three phases is still under preparation. These phases are as follows

1. Phase I: preparation for adoption of REDD+,
2. Phase II: understanding implementation of REDD+,
3. Phase III: full scale result-based implementation of REDD+.

The strategy will be implemented in a coherent and mutually supportive way with the relevant Aichi Biodiversity Targets (5, 7, 11, 14, and 15) and the Strategic Plan for Biodiversity 2011–2020 through various elements of REDD+. The REDD+ strategy plan for India will include mechanism for addressing direct benefit for biodiversity as well as benefit sharing to indigenous and local communities [91].

The incentives so received from REDD+ would be passed to the local communities involved in protection and management of the forests. This will ensure sustained protection of our forests against deforestation [27]. In Indian context, the forest will not be managed for "carbon services" alone, but for all the ecosystem services that are flowing from the forests to the local community. The incentives for carbon services will append to the benefits that the local communities are already receiving from the forest ecosystems [91]. It is estimated that a REDD+ programme for India could provide capture of more than 1 billion tonnes of additional CO_2 over the next 30 years and provide more than USD 3 billion as carbon service incentives under REDD+ [27]. The country specific safeguards will be a part of national strategy with a view to ensure full participation of local communities and all other stakeholders. India's national REDD+ strategy aims to enhance the quantum of forest ecosystem services that flow to the local communities by enhancing and improving the forest and tree cover of the country [91].

Hitherto, India has not prepared a REDD+ strategy as required under the UNFCCC; hence a dedicated institutional structure needs to be put in place to formulate and implement a national REDD+ strategy with a clearly defined mandate, roles, and responsibilities [20]. The issues to be addressed through REDD+ are complex in nature. The implementation of REDD+ frameworks will be strongly influenced by the presence or lack of robust legal framework. The existence of a legal framework that addresses environmental, social and economic issues that reflect the international legal framework of norms and standards is crucial for India to adopt standards on REDD+ [74].

CONCLUSIONS

Currently, there is significant concern over global warming, greenhouse gas emissions, and their possible impacts on society. For successful implementation of REDD+ in India, technological assistance, readiness assistance, and continued political momentum are essential. A fully fledged national REDD+ strategy needs to be developed and implemented along with action plans with an immediate start of pilot activities in selected areas in collaboration of partners at national and international levels. An enormous capacity building at every level of forest hierarchy is needed for successful REDD+ implementation. An appropriate multilateral financial mechanism needs to be developed to operationalize capacity building programmes. The urgency of national REDD+ institutional framework stems from the fact that the implementation of REDD+ will involve tremendous coordination efforts. Pilot projects for REDD+ can be initiated in joint forest management areas. For effective implementation of REDD+ programmes there is need for forests-related legislation in place to enable REDD+ related project development which also includes definitions of ownership, rights, and obligations with respect to the sale and trade of carbon or Payment for Environmental Services (PES) accrued through REDD+ projects.

REFERENCES

1. J. Kishwan, R. Pandey, and V. K. Dadhwal, "Emission removal capability of India's forest and tree cover," Small Scale Forestry, vol. 11, no. 1, pp. 61–72, 2012.
2. C. Streck, R. Sullivan, S. J. Toby, and R. Tarasofsky, Eds., Climate Change and Forestry-Emerging Policy and Market Opportunities, Royal Institute of International Affairs, Brookings Institution Press, Baltimore, Md, USA, 2008.

3. B. G. Mackey, H. Keith, S. L. Berry, and D. B. Lindenmayer, "Green Carbon—the role of natural forests in carbon storage, part 1-A green carbon account of Australia's South-Eastern eucalypt forest, and policy implications," in The Fenner School of Environment & Society, The Australian National University E Press, Australia, 2008.
4. S. L. Brown, P. Schroeder, and J. S. Kern, "Spatial distribution of biomass in forests of the eastern USA,"Forest Ecology and Management, vol. 123, no. 1, pp. 81–90, 1999.
5. J. T. Houghton, Y. Ding, and D. J. Griggs, Eds., Climate Change 2001: The Scientific Basis. Contribution of Working Group I to the Third Assessment Report of the Intergovernmental Panel on Climate Change, Cambridge University Press, Cambridge, UK, 2001.
6. R. A. Houghton, K. T. Lawrence, J. L. Hackler, and S. Brown, "The spatial distribution of forest biomass in the Brazilian Amazon: a comparison of estimates," Global Change Biology, vol. 7, no. 7, pp. 731–746, 2001.
7. C. L. Goodale and M. J. Apps, "Forest sinks in the Northern hemisphere," Ecology Applied, vol. 3, pp. 891–899, 2002.
8. X. Zhang, M. Wang, and X. Liang, "Quantitative classification and carbon density of the forest vegetation in Lüliang Mountains of China," Plant Ecology, vol. 201, no. 1, pp. 1–9, 2009.
9. Food Agriculture Organization, Carbon Sequestration in Dryland Soils, Food and Agriculture Organization of the United Nations, Rome, Italy, 2004.
10. Sir Nicholas Stern, Stern Review: The Economics of Climate Change, Cambridge University Press, Cambridge, UK, 2006.
11. H. J. Albers and E. J. Z. Robinson, "Reducing emissions from deforestation and forest degradation,"Encyclopaedia of Energy, Natural Resources and Environmental Economics, vol. 2, pp. 78–85, 2013.
12. H. Gundimeda, P. Sukhdev, R. K. Sinha, and S. Sanyal, "Natural resource accounting for Indian states—illustrating the case of forest resources," Ecological Economics, vol. 61, no. 4, pp. 635–649, 2007.

13. A. Aggarwal, V. Paul, and S. Das, Forest Resources: Degradation, Livelihoods, and Climate Change, Looking Back to Change Track, TERI, New Delhi, India, 2009.

14. V. K. Bahuguna, K. Mitra, D. Capistrano, and S. Saigal, Root to Canopy: Regenerating Forests Through Community State Partnerships, Winrock International India, Commonwealth Forestry Association, New Delhi, India, 2004.

15. T. Pistorius, C. B. Schmitt, D. Benick, and S. Entenmann, Greening REDD+: Challenges and Opportunities for Forest Biodiversity Conservation, University of Freiburg, Germany, 2010.

16. R. Sud, J. V. Sharma, and A. K. Bansal, International REDD+ Architecture and Its Relevance for India, The Energy and Resources Institute (TERI), New Delhi, India, 2012.

17. V. R. S. Rawat, "REDD Plus in India: from negotiations to implementation," in Proceedings of the Pre-Congress Workshop of 1st Indian Forests Congress, ICFRE, Dehradun, India, 2011.

18. G. R. Van Der Werf, D. C. Morton, R. S. Defries et al., "CO_2 emissions from forest loss," Nature Geoscience, vol. 2, no. 11, pp. 737–738, 2009.

19. EDRI, From RED to REDD+, Africa Regional Dialogue on Forests, Governance & Climate Change, 2010.

20. N. H. Ravindranath, N. Srivastava, I. K. Murthy, S. Malaviya, M. Munsi, and N. Sharma, "Deforestation and forest degradation in India—implications for REDD+," Current Science, vol. 102, pp. 1117–1125, 2012.

21. IISD, Safeguards and Multiple Benefits in a REDD+ Mechanism, International Institute of Sustainable Development, Canada, 2011, www.iisd.org.

22. T. Clements, "Reduced expectations: the political and institutional challenges of REDD+," ORYX, vol. 44, no. 3, pp. 309–310, 2010.

23. M. Collins, E. A. Macdonald, L. Clayton, I. Dunggio, D. W. Macdonald, and E. J. Milner-Gulland, "Wildlife conservation and reduced emissions from deforestation in a case study of Nantu Wildlife Reserve, Sulawesi: 2. An institutional framework for REDD implementation," Environmental Science and Policy, vol. 14, no. 6, pp. 709–718, 2011.

24. S. Engel, S. Pagiola, and S. Wunder, "Designing payments for environmental services in theory and practice: an overview of the issues," Ecological Economics, vol. 65, no. 4, pp. 663–674, 2008.

25. C. McDermott, REDD+ Biodiversity Safeguards: Strength in Diversity—A Study of the Capacity of the UNFCCC to Deliver Equitable, Effective, and Efficient Forest Conservation Policies for Developing Countries, Oxford University School of Geography and the Environment, 2011, Candidate Number: 53172.

26. The Forests Dialogue, Investing in REDD-Plus Consensus Recommendations on Frameworks For the Financing and Implementation of REDD-Plus, TFD Review, A TFD Publication, 2010,www.theforestsdialogue.org.

27. Ministry of Environment and Forests, India's Forests and REDD+, Ministry of Environment and Forests, Government of India.

28. "Global Forest Resources Assessment 2005: Progress towards Sustainable Forest management," FAO Forestry Paper 147, Food and Agriculture Organization of the United Nations, Rome, Italy, 2006.

29. State of Forest Report, Forest Survey of India, Dehradun, India, 2009.

30. B. Pisupati, Safeguarding India's Biological Diversity: The Biological Diversity Act, Farmer's Forum, India's Agriculture Magazine, 2011.

31. ICFRE, Indian Forest Congress, Ministry of Environment and Forests, 2011, http://ify-india.icfre.gov.in/indiaforest.html.

32. Y. Gokhale, Cover Story, Terragreen Magzine, 2010.

33. World Bank, "New global poverty estimates in India," World Bank, 2001,http://www.worldbank.org.in/wbsite/external/countries/southasiaext/indiaextn/0,menupk:295589~pagepk:141159~pipk:141110~thesitepk:295584,00.html.

34. M. Poffenberger, Community and Forest Management in South Asia—A Regional Profile of the Working Group on Community Involvement in Forest Management, Forest, people and policies, IUCN, Switzerland, 2000.

35. World Bank, "An article of World Bank on India: Alleviating Poverty through Forest Development," 2006, World Bank, http://www.worldbank.org/ieg).
36. B. Sinha, C. P. Kala, and A. S. Katiyar, Enhancing Livelihoods of Forest Dependent Communities through Synergizing FDA Activities with Other Development Programmes, Indian Institute of Forest Management, Bhopal, India, 2010.
37. H. G. Champion and V. K. Seth, a Revised Survey of the Forest, Types of India, Government of India, 1968.
38. S. J. Singh, F. Krausmann, S. Gingrich et al., "India's biophysical economy, 1961-2008. Sustainability in a national and global context," Ecological Economics, vol. 76, pp. 60–69, 2012.
39. Forest Survey of India, Demand and Supply of Fuelwood, Timber and Fodder in India, Forest Survey of India, Dehradun, India, 1996.
40. P. P. Singh, "Exploring biodiversity and climate change benefits of community-based forest management," Global Environmental Change, vol. 18, no. 3, pp. 468–478, 2008.
41. R. Pachauri and R. K. Batra, Eds., Directions, Innovations, and Strategies for Harnessing Action for Sustainable Development, Tata Energy Research Institute, New Delhi, India, 2001.
42. National Forestry Action Plan, Ministry of Environment and Forest, Government of India, New Delhi, India, 1999.
43. S. Singh, India: Assessing Management Effectiveness of Wildlife Protected Areas in India, 2001,http://www.iucu.org/themes/forests/protected area/India.pdf.
44. T. N. Khushoo, Environmental Priorities in India and Sustainable Development, Indian Science Congress, Calcutta, India, 1986.
45. F. Warner, Indo Swedish Forestry Programme II: 1982-83 TO, 1986-87, Background Document, Swedish Embassy Dev Cooperation office, New Delhi, India, 1982.
46. J. F. Richards, "Global patterns of land conversion," Environment, vol. 26, no. 9, pp. 6–34, 1984.
47. M. Williams, Forests: The Earth as Transformed by Human Actions, Cambridge University Press, Cambridge, UK,1990.

48. M. Joshi and P. P. Singh, Tropical Deforestation and Forest Degradation: A Case Study from India, XII World Forestry Congress, Quebac City, Canada, 2003.
49. J. Sathaye, K. Andrasko, and P. Chan, "Emissions scenarios, costs, and implementation considerations of REDD-plus programs," Environment and Development Economics, vol. 16, no. 4, pp. 361–380, 2011.
50. Food Agriculture Organization, State of the World's Forests, Food and Agriculture Organization of the United Nations, Rome, Italy, 2007.
51. TERI, Looking Back to Think Ahead: GREEN India 2047, Tata Energy Research Institute, New Delhi, India, 1998.
52. N. H. Ravindranath, R. K. Chaturvedi, and I. K. Murthy, "Forest conservation, afforestation and reforestation in India: Implications for forest carbon stocks," Current Science, vol. 95, no. 2, pp. 216–222, 2008.
53. State of Forest Report, Forest Survey of India, Dehradun, 1987.
54. State of Forest Report, Forest Survey of India, Dehradun, 1989.
55. State of Forest Report, Forest Survey of India, Dehradun, 1991.
56. State of Forest Report, Forest Survey of India, Dehradun, 1993.
57. State of Forest Report, Forest Survey of India, Dehradun, 1995.
58. State of Forest Report, Forest Survey of India, Dehradun, 1997.
59. State of Forest Report, Forest Survey of India, Dehradun, 1999.
60. State of Forest Report, Forest Survey of India, Dehradun, 2001.
61. State of Forest Report, Forest Survey of India, Dehradun, 2003.
62. State of Forest Report, Forest Survey of India, Dehradun, 2005.
63. State of Forest Report, Forest Survey of India, Dehradun, 2011.
64. D. M. Bhat, K. S. Murali, and N. H. Ravindranath, "Formation and recovery of secondary forests in India: a particular reference to Western Ghats in South India," Journal of Tropical Forest Science, vol. 13, no. 4, pp. 601–620, 2001.
65. S. Bajracharya, Community carbon forestry: remote sensing of forest carbon and forest degradation in Nepal [M.S. thesis], International Institute for Geo-information Science and Earth Observation, Enscheda, The Netherlands, 2008.

66. J. O. Niles, S. Brown, J. Pretty, A. Ball, and J. Fay, Potential Mitigation and Income in Developing Countries from Changes in Use and Management of Agricultural and Forest Lands, Center of Environment and Society, Occasional , University of Essex, 2001.

67. United Nations Framework Convention on Climate Change, Submission by India to SBSTA, UNFCCC on Agenda Item 4: Methodological guidance for activities relating to reducing emissions from deforestation and forest degradation and the role of conservation, sustainable management of forests and enhancement of forests carbon stocks in developing countries (UNFCCC Document FCCC/SBSTA/L. 25 dated 3 Dec 2011), 2011.

68. United Nations Framework Convention on Climate Change, The Cancun Agreements: Outcome of the work on the Ad Hoc Working Group on Long-Term Cooperative Action under the Convention. Report of the Conference of the parties on its sixteenth session, held in Cancun from November 29-December 10, 2010. FCCC/CP/2010/7/Add. 1, 2011.

69. Food Agriculture Organization, "Global Forest Resource Assessment 2010," FAO Forestry Paper 163, Rome, Italy, 2010.

70. B. P. Nayak, P. Kohli, and J. V. Sharma, Livelihood of Local Communities and Forest Degradation in India: Issues For REDD+, The Energy and Resources Institute, New Delhi, India, 2012.

71. Ministry of Environment and Forests, Report of the National Forest Commission, Ministry of Environment and Forests, Government of India, New Delhi, India, 2006.

72. P. Davidar, S. Sahoo, P. C. Mammen et al., "Assessing the extent and causes of forest degradation in India: where do we stand?" Biological Conservation, vol. 143, no. 12, pp. 2937–2944, 2010.

73. Ministry of Environment and Forests, "Asia-pacific forestry sector outlook study II: India country report," Working Paper APFSOS II/ WP/2009/06, FAO, Bangkok, Thailand, 2009.

74. N. Moss, R. Nussbaum, and J. Muchemi, REDD+ Safeguards—Background Paper Prepared For REDD+ Partnership, Workshop on Enhancing Coordinated Delivery of REDD+: Emerging Lessons, Best Practices and Challenges, Cancún, Mexico, 2010.

75. D. M. Mwayafu and J. W. Kisekka, Promoting and Implementing REDD+ Safeguards at National Level in East Africa, REDD-net programme, Norwegian Agency for Development Cooperation (NORAD), 2012.
76. T. A. Gardner, N. D. Burgess, N. Aguilar-Amuchastegui et al., "A framework for integrating biodiversity concerns into national REDD+ programmes," Biological Conservation, vol. 154, pp. 61–71, 2012.
77. United Nations Framework Convention on Climate Change, The Cancun Agreements 1/CP. 16, 2010.
78. Greenpeace, REDD: A Common Approach to Safeguards, 2011.
79. United Nations Commission on Human Rights, "Sub-Commission on the Promotion and Protection of Human Rights," in Proceedings of the 22nd session Working Group on Indigenous Populations, New York, NY, USA, 2004.
80. B. Palmer Fry, "Community forest monitoring in REDD+: the 'M' in MRV?" Environmental Science and Policy, vol. 14, no. 2, pp. 181–187, 2011.
81. A. Angelsen, Realising REDD+: National Strategy and Policy Options, Centre for International Forestry Research (CIFOR), Bogor, Indonesia, 2009.
82. A. Angelsen, S. Brown, C. Loisel, L. Peskett, C. Streck, and D. Zarin, Reducing Emissions from Deforestation and Forest Degradation (REDD): An Options Assessment Report, Meridian Institute, Washington, DC, USA, 2009.
83. E. Mattsson, U. M. Persson, M. Ostwald, and S. P. Nissanka, "REDD+ readiness implications for Sri Lanka in terms of reducing deforestation," Journal of Environmental Management, vol. 100, pp. 20–40, 2012.
84. P. Sukhdev, R. Prabhu, P. Kumar et al., REDD+ and a Green Economy: Opportunities for a mutually Supportive relAtionship, UN-REDD Programme Policy Brief, 2012.
85. A. Aggarwal, "Implementation of forest rights act, changing forest landscape, and politics of REDD+ in India," Resources, Energy, and Development, vol. 8, no. 2, pp. 131–148, 2011.
86. Chapter on Climate Change and Forests, Ministry of Environment and Forests, Government of India.

87. Planning Commission, Interim Report of the Expert Group on Low Carbon Strategies for Inclusive Growth, Government of India, 2011.

88. National Action Plan on Climate Change, Prime Minister's Council on Climate Change, Government of India, undated, http://pmindia.nic.in/Pg01-52.pdf.

89. J. Eliasch, Climate Change: Financing Global Forests, Office of Climate Change, 2008,http://www.illegal logging.info/sites/default/files/uploads/Fullreporteliaschreview1.pdf.

90. C. A. Harvey, B. Dickson, and C. Kormos, "Opportunities for achieving biodiversity conservation through REDD," in Conservation Letters, pp. 1–9, Wiley Periodicals, 2009.

91. United Nations Framework Convention on Climate Change, Inputs of Climate change Division on Agenda Item 7 of subsidiary body on scientific, technical and technological advice for Sixteenth meeting at Montreal, 30 April 5 May 2012, 2012.

Chapter 4

Connecting Soil Organic Carbon and Root Biomass with Land-Use and Vegetation in Temperate Grassland

Devan Allen McGranahan[1], Aaron L. Daigh[1], Jessica J. Veenstra[2], David M. Engle[3], James R. Miller[4], and Diane M. Debinski[5]

[1]School of Natural Resource Sciences Range Science Program, North Dakota State University, Fargo, ND 58108-6050, USA

[2]Natural Sciences, Flagler College, St. Augustine, FL 32085-1027, USA

[3]Department of Natural Resource Ecology and Management, Oklahoma State University, Stillwater, OK 74078-6013, USA

[4]Natural Resources and Environmental Sciences, University of Illinois, Urbana, IL 61801, USA

[5]Department of Ecology, Evolution, and Organismal Ecology, Iowa State University, Ames, IA 50011, USA

ABSTRACT

Soils contain much of Earth's terrestrial organic carbon but are sensitive to land-use. Rangelands are important to carbon dynamics and are among ecosystems most widely impacted by land-use. While common practices like grazing, fire, and tillage affect soil properties directly related to soil carbon dynamics, their magnitude and direction of change vary among ecosystems and with intensity of disturbance. We describe variability in soil organic carbon (SOC) and root biomass—sampled from 0–170 cm and 0–100 cm, respectively—in terms of soil properties, land-use history, current management, and plant community composition using linear regression and multivariate ordination. Despite consistency in average values of SOC and root biomass between our data and data from rangelands worldwide, broad ranges in root biomass and SOC in our data suggest these variables are affected by other site-specific factors. Pastures with a recent history of severe grazing had reduced root biomass and greater bulk density. Ordination suggests greater exotic species richness is associated with lower root biomass but the relationship was not apparent when an invasive species of management concern was specifically tested. We discuss how unexplained variability in belowground properties can complicate measurement and prediction of ecosystem processes such as carbon sequestration.

INTRODUCTION

Soils constitute the greatest stock of terrestrial organic carbon [1] and soil properties can be affected by land-use and management [2, 3]. Globally, approximately one-quarter of the potential carbon sequestration in soils occurs in rangelands [4]. Rangelands are also one of the most widespread human-impacted biomes on Earth [5], making their role in carbon sequestration sensitive to land-use and climate change [6, 7].

The effects of grazing and fire, specifically, can be varied and opposing. In rangeland, grazing can increase bulk density [8, 9] and has a neutral or negative effect on soil organic carbon (SOC) [10–12]. Likewise, grazing can either increase or decrease root production

[13, 14]. Fire increases root growth in tallgrass prairie [13, 15], but across fire-adapted ecosystems the effect of fire on soil carbon varies with severity and temporal scale [16–19]. Temporal scale is especially important to SOC because although SOC can decline over just a few years, SOC accumulation occurs on the scale of decades [20, 21].

Vegetation affects soil carbon stocks by depositing organic matter in the soil. Two major pathways of organic matter input—root tissue and exudates—directly involve plants [22]. Roots contribute to SOC pools through rhizodeposition [23] and the longer residence time of carbon from root tissue than shoot tissue [24]. Abiotic factors can also affect root growth and SOC. Soil clay content has been associated with greater SOC in many soils [25–27]. Bulk density can limit root growth and decrease SOC [28, 29].

Vegetation changes that affect organic matter input deep in the soil profile have an important impact on carbon dynamics. For instance, invasive deep-rooted, warm-season (C_4) grasses can increase carbon sequestration by increasing organic matter deposition deep in the soil profile [30]. Conversely, the invasion of shallow-rooted, cool-season (C_3) grasses might reduce soil carbon because cool-season grasses contribute less root tissue than native warm-season grasses deep in the soil profile [31, 32] despite greater root biomass near the soil surface [33].

Carbon dynamics deep in the soil profile are especially important but infrequently studied. Increasing the considered depth from 100 cm to 300 cm increases the global SOC budget by 56% [34]. Carbon deeper in the soil profile is subject to different soil structure, chemistry, and biotic activity, which might contribute to greater carbon sequestration [35].

We studied the effect of grazing history, prescribed fire, and the invasion of an exotic C_3grass on root biomass and SOC in rangeland managed with fire and grazing in the tallgrass prairie region of central North America. We describe variability in SOC and root biomass—sampled from 0–170 cm and 0–100 cm, respectively—in terms of soil properties, land-use history, current management, and plant community composition using linear regression and multivariate ordination methods. We expected pastures with a recent history of severe grazing to have less root biomass and lower SOC. We also expected plots with greater abundance of exotic C_3 invasive species to have less root biomass. Finally, we expected to associate belowground properties with variation in aboveground plant community composition.

METHODS

Study Location and Site History

Our research was conducted in conjunction with ongoing research in the Grand River Grasslands, a 30,000 ha working landscape in Ringgold County, IA, and Harrison County, MO [36]. Study tracts were initially identified as having medium to high potential for prairie conservation and restoration, based on the observed presence of native species indicating parcels of remnant prairie (The Nature Conservancy, unpublished data). When the Grand River Grasslands research project began in 2006, a pretreatment vegetation survey of potentially-remnant tracts confirmed a high incidence of native plant species as well as a range of invasion by nonnative plants [37].

Tracts were identified by historical grazing management. Grazing histories were reconstructed through interviews with current and former managers; four of the tracts were reported ungrazed for at least six years prior to the beginning of the study while five had been grazed by cattle (Bos taurus) at high stocking rates (ca. 15 animal unit months/ha) (Table 1) [37, 38]. At the time of this study, tracts were assigned to treatments for the purposes of a fire and grazing experiment that divided the tracts into moderately grazed and ungrazed [39]. We did not expect these recent changes in management to affect belowground soil and root properties (although the effect was tested as part of multivariate analyses, see below). However, because fire has been shown to reduce belowground biomass in tallgrass prairie, for example [15], we did record time-since-fire for each patch for inclusion in our analyses.

Table 1: Summary of historical (2000–2006) and 2010 grazing management information and tall fescue abundance (as percent canopy cover) for six pastures in the Grand River Grasslands of Ringgold County, IA, and Harrison County, MO, USA Severe grazing refers to approximately 15 animal unit months/ha [37]

Pasture	Grazing		Tall fescue canopy cover (%)		
	Historical	Current	Minimum	Mean (±se)	Maximum
Lee Trail	Not recently grazed	Moderately grazed	13	35 (±8)	61
Pawnee	Not recently grazed	Ungrazed	0	<1	<1
Pyland North	Recently severely grazed	Moderately grazed	38	59 (±7)	75
Pyland South	Recently severely grazed	Moderately grazed	20	37 (±5)	53
Pyland West	Recently severely grazed	Moderately grazed	17	50 (±8)	63
Ringgold North	Not recently grazed	Ungrazed	0	<1	1

Sample Collection and Analysis

As part of ongoing research in these tracts, six modified Whittaker plots per pasture were located with respect to soil series as described by McGranahan et al. [37, 40]. All study tracts were classified to the Gara-Armstrong-Pershing association [41]. Two soil series—Gara loam and Armstrong loam (Fine-loamy, mixed, superactive, mesic Mollic Hapludalf, parent material: glacial till; and Fine, smectitic, mesic Aquertic Hapludalf, parent material: loess over palesol formed in glacial till, respectively [41])—dominated the study tracts, and within each tract three plots were located within each soil series. Slopes spanned three classifications (C, D, and E) and ranged from 8 to 35%.

In June 2010, we located 31, 500 m^2 modified Whittaker plots (permanently-located vegetation survey plots established in 2006 for estimation of canopy cover [37, 42]) across six tracts in the Grand River Grasslands. Although each tract has six permanent plots, some plots were not accessible by the heavy equipment required for soil sampling. At each plot, we sampled the abundance of tall fescue by recording canopy coverage from 10, 0.5 m^2 quadrats according to the Daubenmire [43] canopy cover index. We used the mean canopy cover of these 10 quadrats to represent tall fescue abundance in our analyses.

We extracted four adjacent 7.5 cm diameter soil cores from the approximate center of each vegetation plot with a vehicle-mounted hydraulic Giddings probe. Three, 100 cm cores were analyzed for root biomass at 20 cm intervals to determine rooting depth. Twenty-centimeter sections of each core were soaked overnight in a 1% solution of sodium hexametaphosphate (Calgon) [14]. We separated root tissue from soil particles with a sieve and bucket arrangement similar to Lauenroth and Whitman [44]: water was flushed through each soaked core section in a 10 mesh sieve mounted atop a 19L bucket. Mineral material sank in the bucket while root tissue that passed through the 10 mesh sieve floated and was collected in a 40 mesh sieve. Root tissue was collected with tweezers from remaining particles in the sieve and dried for 48 hours at 45°C, with root biomass expressed as mass per unit area [31, 32].

The fourth soil core was sampled to 170 cm and analyzed for soil organic carbon (SOC) at varying depth intervals: 10 cm intervals,

0–60 cm; 20 cm intervals, 60–140 cm, and a 30 cm interval from 140 to 170 cm. To standardize depth intervals when making comparisons with root biomass—which was sampled 0–100 cm in 20 cm increments—we summed 10 cm incremental SOC data from the top 0–60 cm into 20 cm increments. These cores were air-dried and stored unsealed.

Prior to laboratory analysis, the fourth core from each plot was also analyzed for variables that might affect root penetration, including clay content and bulk density in 20 cm increments, depth to argillic and gleyed horizons (which indicate clay accumulation and anoxic conditions, resp.), and depth to an observable plow layer. Based on previous experience with these similar soil series, A. Daigh identified potentially root-limiting thresholds for clay content (27%) and bulk density (1.4 g/cm^3) and for each core determined the depth at which the root-limiting layer was first observed.

Because our study considered only organic carbon, we tested for and eliminated any inorganic carbon fraction from soil samples. Total carbon and inorganic carbon was determined by the Iowa State University Plant and Soil Analysis Lab, Ames, IA, USA, using the dry combustion and modified pressure calcimeter methods. pH was determined with a glass electrode in a 1 : 1 soil to water suspension. Composite samples from each depth interval were analyzed for total carbon and pH. For samples <7.0 pH, total carbon was assumed to equal organic carbon. Samples ≥7.0 pH were reanalyzed for inorganic carbon, which, when subtracted from total carbon, gives the organic carbon fraction. Prior to submission to the lab, we determined soil bulk density via the soil core method to calculate soil organic carbon on a volumetric rather than gravimetric basis, the standard for reporting and comparing soil carbon data globally, for example [34].

Data Analysis

Belowground Properties and Land-Use

To determine the relationship between soil variables (root mass, SOC, and bulk density) and land-use variables (grazing history, time-since-fire, and tall fescue cover) we constructed linear mixed effect regression (LME) models with the lmer function in the lme4 package (version 1.0-5) for the R statistical environment (version 3.0.2) [45, 46]. Response variables included both whole core root biomass (0–100 cm) and surface biomass (0–20 cm); whole core SOC (0–170 cm), surface SOC (0–20 cm), and surface percent SOC (0–10 cm); and surface bulk density (0–20 cm). We also compared total SOC, percent SOC, and root mass in the top 0–20 cm against clay content and bulk density (0–20 cm) and tested for a correlation between bulk density and clay content.

In lieu of P values as a measure of statistical significance, we estimated 95% confidence intervals for grazing history and tall fescue cover using the simulation method developed by Nakagawa and Cuthill [47], which compares 1000 simulations of the LME model to empirical response variable data. To test the goodness-of-fit of the mixed-effect regression model we calculated a coefficient of determination (R^2) with a custom rsquared.lme function following Nakagawa and Schielzeth [48]. The rsquared.lme function extracts variance components from the lme model and calculates marginal R^2 values that represent the goodness-of-fit for the fixed-effect term.

Multivariate Analysis of Soil, Root, and Vegetation Data

We used ordination to identify patterns among root mass, SOC, and soil properties data. We performed a Principal Components Analysis (PCA) with an unconstrained model using the rda function in the vegan package (version 2.0-7) for the R statistical environment [49]. We tested several physical and management-related factors against the ordination using the envfit function in vegan; these variables included tall grazing history, tall fescue cover, current grazing treatment, soil series, and slope of sampled plot.

We also used ordination to test for association between plant community composition and variation in soil and root properties. First, we extracted the principal components (PCs) from the PCA of soil and root data. Together, the PCs represent composite variables that each account for a proportion of variation in the root/soil dataset; as PCs are added the proportion of variation explained by the composite variables accumulates. We sought to include as many PCs as necessary to account for at least 70% of variation in the soil/root data.

Second, we set the PCs as constraints in a Constrained Analysis of Proximities (function capscale in vegan), which performs a constrained ordination based on a user-defined distance metric and is similar to the unconstrained ordination Multi-Dimensional Scaling. For plant community data we used the 2006 pretreatment vegetation survey [37] and used the Canberra distance metric. As applied here, the constrained ordination first describes variation in plant community composition along defined axes—in this case, the composite variables of soil/root data represented by the PCs—then proceeds to explain remaining variation via unconstrained ordination. To determine how useful the soil/root composite variables were in describing variation in plant community composition, we compared an unconstrained ordination of the vegetation data to the constrained ordination model using the anova function.

RESULTS AND DISCUSSION

Relationship between Belowground Properties and Depth

Root biomass and SOC decreased rapidly with depth (Figure 1). Bulk density generally increased with depth; wide variation in samples from the 60 to 80 cm increment (Figure1(c)) reflects a low-density sand lens (likely an isolated/local variation in the glacial till) in a subset of samples. Across all plots, the 0–20 cm increment of the soil profile contained the greatest root biomass and SOC (Table 2). In these soils, sampling to 100 cm increased SOC stock and root biomass by an average of 45% and 40%, respectively, when compared to amounts contained in the top 20 cm of the profile. For SOC, extending the sampling depth to 170 cm increased the estimate of SOC stock in the profile by an average of 63%. Deep sampling clearly increases the amount of organic carbon and root biomass accounted for under these grasslands.

Table 2: Descriptive statistics for soil organic carbon (SOC) and root biomass from 31 vegetation plots across five study tracts in the Grand River Grasslands of Ringgold County, IA, and Harrison County, MO, USA

Variable	Minimum	Mean	Maximum
Total SOC in 170 cm profile (kg/m^2)	11.4	16.8	27.4
Percent of total SOC in top 20 cm	21.5%	37.2%	51.4%
Percent of total SOC in top 40 cm	28.8%	56.9%	71.5%
Percent of total SOC in top 100 cm	38.0%	81.9%	96.9%
Total root biomass in 100 cm profile (mg/cm^2)	53.6	102.0	212.4
Percent of total root biomass in top 20 cm	33.1%	69.8%	96.5%
Percent of total root biomass in top 40 cm	52.8%	80.5%	98.0%

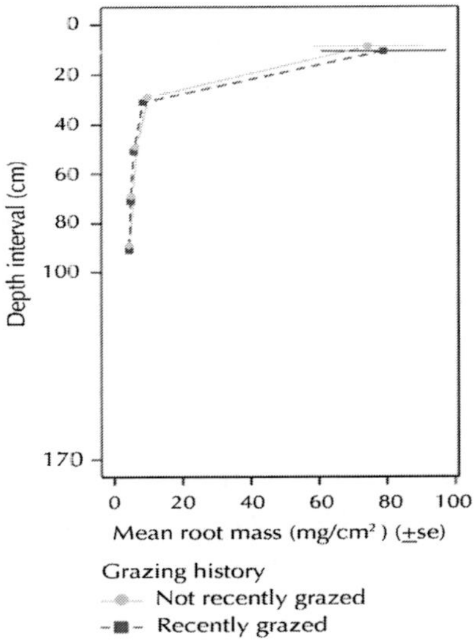

(a)

Figure 1: Root biomass, soil organic carbon (SOC), and bulk density for six tallgrass prairie pastures in the Grand River Grasslands of Ringgold County,

IA, and Harrison County, MO, USA, plotted by depth (0–170 cm) and grazing history. See Methods for sampling procedures including depth intervals.

In terms of vertical distribution, root biomass and SOC tended to be concentrated in the upper portion of the soil profile, a pattern consistent with other work. On average, we found 81% of root biomass in the top 40 cm of the soil profile (Table 1), similar to the 83% average for the top 30 cm of temperate grassland worldwide [50]. In another global review, Jobbágy and Jackson [34] report 70% of root biomass concentrated in the 0–20 cm increment and an additional 17% in the 20–40 cm increment, which is congruous with our mean values of 70% and 11%, respectively. Regarding SOC, Jobbágy and Jackson [34] report an average total of 16 kg/m^2 for the top 200 cm in temperate grasslands worldwide, very near the 17 kg/m^2 mean reported here. Broken down by depth increments, Jobbágy and Jackson [34] report an average of 42% of SOC distributed in the 0–20 cm increment and 23% in the 20–40 cm increment, again consistent with our values of 37% and 20%, respectively.

Land-Use versus Natural Variation in Belowground Properties

Our data indicate that previous grazing management has affected both belowground and aboveground properties of these grasslands. Pastures with a recent history of severe grazing had lower root biomass in both the top 20 cm and the entire 100 cm profile than other pastures, and greater bulk density (Table 3). These results indicate both biotic and abiotic effects of soil compaction associated with intensive livestock management [11, 13,51]. That these effects persist at least five years following the cessation of severe grazing highlights the influence of land-use legacies on biophysical properties and ecological pattern and process [52, 53]. Likewise, these recently severely-grazed pastures have different plant community composition, including lower native species richness and a greater abundance of tall fescue [37, 40]. But tall fescue abundance was associated with neither SOC nor root biomass (Table 3). Likewise, current management showed no effect on belowground properties. Specifically, prescribed fire had no association with root biomass or SOC in the top 20 cm of the soil profile (95% CI = −2.6–0.7, R^2=0.05 and 95% CI = −1407–225, R^2=0.06, resp.).

Table 3: Results of six multiple linear mixed effect regression models each comparing response variables against tall fescue abundance and grazing history as independent variables (fixed effects). Lower and upper bounds define simulated 95% confidence intervals while reports goodness-of-fit for multiple fixed effects

Response variable	Tall fescue		Grazing history		R^2
	Lower	Upper	Lower	Upper	
Soil organic carbon					
0–20 cm	−2.9	20.6	−927	316	0.07
0–100 cm	−58	115	−5339	3307	0.01
% 0–10 cm	−0.02	0.02	−1.68	0.51	0.10
Root mass					
0–20 cm	−0.01	0.06	−4.1	−0.9	0.24
0–100 cm	−0.01	0.18	−13.7	−3.6	0.28
Bulk density					
0–20 cm	−0.002	0.002	0.02	0.2	0.22

We observed several properties in the upper portion of the soil column that could physically impair root penetration. The average depth of the shallowest root-limiting layer was 12 (±2) cm, and it appears to have an association with reduced root biomass in the top 20 cm of the soil profile (95% CI = 0.01–0.11, R^2=0.13), but not percent SOC (95% CI = −0.02–0.04, R^2=0.01). There was no correlation between bulk density and clay content in the top 20 cm, and neither had an association with SOC or root biomass at the same depth (Table 4), contrary to predictions [29].

Table 4: Results of four multiple linear mixed effect regression models each comparing response variables against bulk density and clay content as independent variables (fixed effects). Lower and upper bounds define simulated 95% confidence intervals while R^2 reports goodness-of-fit for multiple fixed effects

Response variable	Bulk density		Clay content		R^2
	Lower	**Upper**	**Lower**	**Upper**	
Soil organic carbon					
SOC 0–20 cm	−2052	1857	−33.5	15.5	0.01
% 0–10 cm	−4.8	0.7	−0.03	0.04	0.07
Root mass (0–20 cm)	−10.8	2.2	−0.09	0.06	0.06
Bulk density (0–20 cm)			−0.006	0.002	0.03

Nine plots on four of the six pastures showed evidence of a plow layer, observed at an average depth of 7.7 (±1.5) cm. These data suggest that cultivation had occurred at some point since European settlement of the area in the late 19th century, contrary to our previous efforts to document agricultural histories (interviews with managers and searches through local US Department of Agriculture records) that presented no evidence of tillage. While historical cultivation is not surprising and is in fact expected, it is clear that even land-use activity that occurred long ago and/or briefly in time can leave a long-lasting imprint on soil. But the long-term impact of such activity is less clear: the presence of a plow layer was not associated with differences in bulk density (95% CI = −0.12–0.07, R^2=0.01) or amount of SOC (95% CI = −284–1005, R^2=0.03) in the top 20 cm of soil profiles. For these pastures, it is possible that tillage occurred so long ago that soil has since recovered [20].

Multivariate Analyses of Belowground Properties and Vegetation

Although the ordination of soil and root data highlighted several patterns among belowground properties, we did not observe patterns that indicate associations between belowground properties, land-use, and vegetation. The PCA revealed three general trends in variation within the soil and root data along which correlated variables clustered (Figure 2): one following root biomass, another SOC, and a third combining clay content and bulk density. Variables related to SOC loaded most heavily along PC1 while variables related to root biomass contributed in greater proportion to PC2 (Table 5). The first three axes of the PCA accounted for 72% of the variation in belowground properties, but there was no association between the PCA and factors that might explain variation in belowground properties, including historical and current grazing, tall fescue abundance, soil series, and slope (P>0.1). We did not expect current management to affect either root biomass or SOC given the low severity and brief time span of the experimental disturbance regime [18, 54].

Table 5: Loadings along first three Principal Components (PC1, PC2, and PC3) for 13 belowground properties measured from six pastures in the Grand River Grasslands of Ringgold County, IA, and Harrison County, MO, USA. Plotting code refers to plotted text in Figure 2

Variable type	Plotting code	Meaning	PC1	PC2	PC3
Soil organic carbon (SOC)	Total SOC	Total mass of SOC in 0–100 cm column	0.76	−0.54	−0.5
	SOC 0–10 cm	SOC mass in 0–10 cm	0.98	−0.33	0.63
	SOC 0–20 cm	SOC mass in 0–20 cm	0.79	−0.37	−0.68
	% total SOC 0–10 cm	Proportion of total SOC mass in 0–10 cm	0.98	−0.31	0.63
	% total SOC 0–20 cm	Proportion of total SOC mass in 0–20 cm	1.13	−0.43	0.13
	% SOC 0–20 cm	% SOC, 0–20 cm	1.12	−0.35	0.12
Root mass	Total root mass	Total root biomass in 0–100 cm column	0.73	0.9	0.07
	Mean root mass 20 cm	Root biomass in 0–20 cm	0.77	0.91	0.07
	% total root mass in 0–20 cm	Proportion of total root biomass in 0–20 cm	0.7	0.81	0.13

Root limitations	Mean clay content 0–20 cm	Average clay content of 0–10 and 11–20 cm intervals	−0.11	−0.22	0.51
	Bulk density 0–10 cm	Bulk density 0–10 cm	−0.43	−0.34	0.21
	Shallowest layer	Shallowest depth (cm) of a soil property expected to limit root penetration[1]	0.38	0.47	−0.29
	A depth	Depth of a horizon (topsoil)	0.81	−0.09	−0.66

[1]Variables used to determine limitation to rooting depth: clay content = 27%, bulk density = 1.4, observation of argillic or gleyed horizon, or a plow layer.

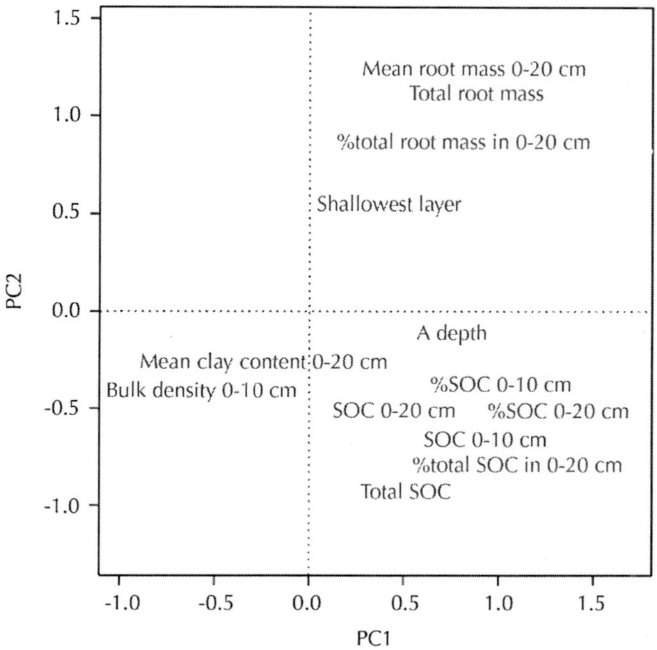

Figure 2: Relationships among 13 belowground properties with respect to first two Principal Components from Principal Components Analysis. For description of plotted text codes and quantified loadings for each variable, see Table 4.

Not surprisingly—given the concentration of root biomass and SOC near the soil surface (Table 2)—total root biomass (0–100 cm) and total SOC (0–170 cm) each clustered with their respective shallow measures (0–20 cm). These results suggest that unless a complete accounting of the carbon budget is required, shallow sampling is sufficient to characterize root and SOC dynamics through much of the soil column.

An unexpected result of the multivariate analysis is the relationship between root biomass and SOC. Whereas we expected a direct, linear correlation between these two variables, they in fact occur orthogonal to each other in ordination space (Figure 2). Both root biomass and SOC variables appear to have a negative relationship with clay content and bulk density, although no associations were indicated by linear regression models (Table4). The ordination supports above evidence that deeper root-limiting layers are associated with greater root biomass and suggests that deeper topsoil (A horizon) might tend to have greater root biomass and SOC.

There is evidence that plant community composition is associated with variation in belowground properties. PC1 and PC2—composite variables from the PCA ordination of multivariate soil and root data most influenced by 0–10 cm SOC and 0–20 cm root biomass, respectively (Table 5)—were significant terms in the constrained ordination of plant community composition (PC1: P=0.01, PC2: P=0.04) Likewise, the first axis of the constrained ordination was significantly associated with variation in plant community composition (CAP1: P=0.005). There appears to be an association between PC2 and exotic species richness in ordination space (Figure 3); although PC2 is the axis of secondary variation in the PCA, it is also the axis most influenced by root biomass. Thus, these data might suggest a connection between greater exotic species richness and reduced root biomass, which supports a pattern seen elsewhere in tallgrass prairie documenting greater root biomass among native species versus exotic species [32]. But the three constraining axes explained just 13% of variation in the ordination of vegetation data, leaving substantial amount of variation in plant community composition unexplained by belowground properties. As shown elsewhere, grazing history, tall fescue abundance, and the ratio of native to exotic species strongly influence plant community composition in these grasslands [37, 39].

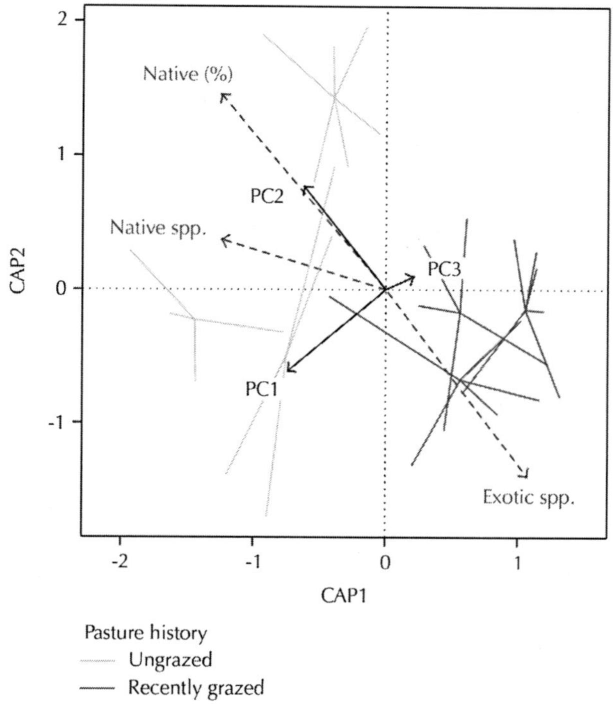

Figure 3: Results of a constrained ordination of plant community composition using composite variables of belowground properties created from soil and root PCA. Spiderplots group sampled modified Whittaker plots by grazing history, thick black arrows show constraining variables, and thin red arrows overlay plant community metadata fitted to the ordination ("Native (%)" = proportion of native species in community, "Native spp." = native species richness, and "Exotic spp." = exotic species richness).

Land-Use, Variability, and Ecosystem Processes

Given that sampling deeper in the soil profile makes important contributions to the global carbon cycle [34], we sampled soil organic carbon (SOC) and root biomass to 170 cm and 100 cm, respectively, in temperate rangeland soils. Although on one hand such deep sampling did increase the amount of soil organic carbon we can account for in this system, on the other hand our ability to relate these data to ecosystem-level processes like carbon sequestration is limited by (i) the high degree of variability we observed in belowground properties at shallow soil layers and (ii) the lack of pattern between variability in belowground properties and known land-uses and plant composition.

Most relevant to carbon budgets, we did not observe any relationship between vegetation and SOC or root biomass at any level of the soil profile, let alone at depth. Although the majority of organic matter is found in the upper strata of the soil, carbon deep in the soil profile might have a disproportionately greater influence on the long-term carbon cycle if deep carbon has a longer residence time than carbon higher in the profile, for example [55]. Other studies indicate exotic plants can affect the vertical distribution of root biomass in native communities [30–32], but greater root biomass in native grass versus tall fescue stands does not necessarily lead to increased SOC [56]. Likewise, our results do not suggest that tall fescue, specifically, affects root biomass or SOC in these rangelands at either the fine or broad temporal scales considered here, despite indications of a community-level association between exotic species richness and reduced root biomass (Figure 3). If one were to consider the gradient of invasion/floristic degradation across the study tracts as a space-for-time substitution, for example [57], one could draw two conclusions: either invasive species might not reduce the carbon sequestration potential of these rangelands or the restoration of native plant dominance does not seem likely to increase carbon sequestration potential of invaded/degraded rangeland in the foreseeable future.

CONCLUSIONS

Although sampling deep into the soil profile substantially increased estimated stocks of SOC and root biomass in these rangelands, observed dynamics between soil properties, management, and plant communities appear restricted to the upper 20 cm of the soil profile. These results corroborate two important themes in soil carbon research. First, increasing the sampling depth contributes to substantially greater soil organic carbon (SOC) stocks in temperate grassland soils. Second, vegetation, land-use management, and soil properties interact to affect soil carbon and root biomass stocks, but this relationship is not necessarily straightforward. Together, these results support the claim that rangeland soils are important carbon pools but also suggest it is unlikely that rangeland plant communities can be effectively categorized by their carbon sequestration potential without considering biotic and abiotic factors, even within climate zones and regions.

Conflict of Interests

The authors declare that there is no conflict of interests regarding the publication of this paper.

ACKNOWLEDGMENTS

The authors thank the Leopold Center for Sustainable Agriculture at Iowa State University, Ames, Iowa, and the US Environmental Protection Agency's P3 Award for funding data collection. Analysis was supported by the Iowa State and Oklahoma State Agriculture Experiment Stations. The authors appreciate the logistical contributions of R. Harr, R. Limb, and M. Kirkwood.

REFERENCES

1. N. H. Batjes, "Total carbon and nitrogen in the soils of the world," European Journal of Soil Science, vol. 47, no. 2, pp. 151–163, 1996.
2. A. P. da Silva, B. D. Kay, and E. Perfect, "Management versus inherent soil properties effects on bulk density and relative compaction," Soil and Tillage Research, vol. 44, no. 1-2, pp. 81–93, 1997.
3. R. L. McCulley, I. C. Burke, J. A. Nelson, W. K. Lauenroth, A. K. Knapp, and E. F. Kelly, "Regional patterns in carbon cycling across the Great Plains of North America," Ecosystems, vol. 8, no. 1, pp. 106–121, 2005.
4. R. F. Follett and D. A. Reed, "Soil carbon sequestration in grazing lands: societal benefits and policy implications," Rangeland Ecology and Management, vol. 63, no. 1, pp. 4–15, 2010.
5. E. C. Ellis and N. Ramankutty, "Putting people in the map: anthropogenic biomes of the world," Frontiers in Ecology and the Environment, vol. 6, no. 8, pp. 439–447, 2008

6. R. T. Conant and K. Paustian, "Potential soil carbon sequestration in overgrazed grassland ecosystems," Global Biogeochemical Cycles, vol. 16, no. 4, pp. 90-1–90-9, 2002
7. C. Dean, G. W. Wardell-Johnson, and R. J. Harper, "Carbon management of commercial rangelands in Australia: major pools and fluxes," Agriculture, Ecosystems & Environment, vol. 148, pp. 44–64, 2012
8. J. T. Manley, G. E. Schuman, J. D. Reeder, and R. H. Hart, "Rangeland soil carbon and nitrogen responses to grazing," Journal of Soil and Water Conservation, vol. 50, no. 3, pp. 294–298, 1995
9. G. Pineiro, J. M. Paruelo, M. Oesterheld, and E. G. Jobbágy, "Pathways of grazing effects on soil organic carbon and nitrogen," Rangeland Ecology and Management, vol. 63, no. 1, pp. 109–119, 2010.
10. S. Bagchi and M. E. Ritchie, "Introduced grazers can restrict potential soil carbon sequestration through impacts on plant community composition," Ecology Letters, vol. 13, no. 8, pp. 959–968, 2010
11. N. P. He, Y. H. Zhang, Q. Yu et al., "Grazing intensity impacts soil carbon and nitrogen storage of continental steppe," Ecosphere, vol. 2, no. 1, article 8, 2011.
12. E. Medina-Roldán, J. Paz-Ferreiro, and R. D. Bardgett, "Grazing exclusion affects soil and plant communities, but has no impact on soil carbon storage in an upland grassland," Agriculture, Ecosystems and Environment, vol. 149, pp. 118–123, 2012.
13. L. C. Johnson and J. R. Matchett, "Fire and grazing regulate belowground processes in tallgrass prairie," Ecology, vol. 82, no. 12, pp. 3377–3389, 2001
14. E. Pucheta, I. Bonamici, M. Cabido, and S. Díaz, "Below-ground biomass and productivity of a grazed site and a neighbouring ungrazed exclosure in a grassland in central Argentina," Austral Ecology, vol. 29, no. 2, pp. 201–208, 2004.
15. R. F. Limb, S. D. Fuhlendorf, D. M. Engle, and J. D. Kerby, "Growing-season disturbance in tallgrass prairie: evaluating fire and grazing on Schizachyrium scoparium," Rangeland Ecology and Management, vol. 64, no. 1, pp. 28–36, 2011.

16. D. G. Neary, C. C. Klopatek, L. F. DeBano, and P. F. Ffolliott, "Fire effects on belowground sustainability: a review and synthesis," Forest Ecology and Management, vol. 122, no. 1-2, pp. 51–71, 1999

17. M. I. Bird, E. M. Veenendaal, C. Moyo, J. Lloyd, and P. Frost, "Effect of fire and soil texture on soil carbon in a sub-humid savanna (Matopos, Zimbabwe)," Geoderma, vol. 94, no. 1, pp. 71–90, 2000

18. R. W. S. Fynn, R. J. Haynes, and T. G. O'Connor, "Burning causes long-term changes in soil organic matter content of a South African grassland," Soil Biology and Biochemistry, vol. 35, no. 5, pp. 677–687, 2003.

19. H. Knicker, "How does fire affect the nature and stability of soil organic nitrogen and carbon? A review," Biogeochemistry, vol. 85, no. 1, pp. 91–118, 2007.

20. S. G. Baer, D. J. Kitchen, J. M. Blair, and C. W. Rice, "Changes in ecosystem structure and function along a chronosequence of restored grasslands," Ecological Applications, vol. 12, no. 6, pp. 1688–1701, 2002

21. S. G. Baer, C. K. Meyer, E. M. Bach, R. P. Klopf, and J. Six, "Contrasting ecosystem recovery on two soil textures: implications for carbon mitigation and grassland conservation," Ecosphere, vol. 1, no. 1, article 5, 2010

22. C. Rumpel and I. Kögel-Knabner, "Deep soil organic matter—a key but poorly understood component of terrestrial C cycle," Plant and Soil, vol. 338, no. 1-2, pp. 143–158, 2011

23. A. R. Wilts, D. C. Reicosky, R. R. Allmaras, and C. E. Clapp, "Long-term corn residue effects: harvest alternatives, soil carbon turnover, and root-derived carbon,"Soil Science Society of America Journal, vol. 68, no. 4, pp. 1342–1351, 2004

24. D. P. Rasse, C. Rumpel, and M.-F. Dignac, "Is soil carbon mostly root carbon? Mechanisms for a specific stabilisation," Plant and Soil, vol. 269, no. 1-2, pp. 341–356, 2005

25. P. Sollins, P. Homann, and B. A. Caldwell, "Stabilization and destabilization of soil organic matter: mechanisms and controls," Geoderma, vol. 74, no. 1-2, pp. 65–105, 1996.

26. R. Alvarez and R. S. Lavado, "Climate, organic matter and clay

content relationships in the Pampa and Chaco soils, Argentina," Geoderma, vol. 83, no. 1-2, pp. 127–141, 1998.

27. J. Leifeld, S. Bassin, and J. Fuhrer, "Carbon stocks in Swiss agricultural soils predicted by land-use, soil characteristics, and altitude," Agriculture, Ecosystems and Environment, vol. 105, no. 1-2, pp. 255–266, 2005
28. R. Lal and J. M. Kimble, "Importance of soil bulk density and methods of its importance," in Assessment Methods for Soil Carbon, Lewis, Boca Raton, Fla, USA, 2001.
29. K. R. Brye, C. P. West, and E. E. Gbur, "Soil quality differences under native tallgrass prairie across a climosequence in Arkansas," The American Midland Naturalist, vol. 152, no. 2, pp. 214–230, 2004
30. M. J. Fisher, I. M. Rao, M. A. Ayarza et al., "Carbon storage by introduced deep-rooted grasses in the South American savannas," Nature, vol. 371, no. 6494, pp. 236–238, 1994
31. A. Tufekcioglu, J. W. Raich, T. M. Isenhart, and R. C. Schultz, "Fine root dynamics, coarse root biomass, root distribution, and soil respiration in a multispecies riparian buffer in Central Iowa, USA," Agroforestry Systems, vol. 44, no. 2-3, pp. 163–174, 1998
32. B. J. Wilsey and H. W. Polley, "Aboveground productivity and root-shoot allocation differ between native and introduced grass species," Oecologia, vol. 150, no. 2, pp. 300–309, 2006
33. K. A. Fink and S. D. Wilson, "Bromus inermis invasion of a native grassland: diversity and resource reduction," Botany, vol. 89, no. 3, pp. 157–164, 2011.
34. E. G. Jobbágy and R. B. Jackson, "The vertical distribution of soil organic carbon and its relation to climate and vegetation," Ecological Applications, vol. 10, no. 2, pp. 423–436, 2000
35. H. Blanco-Canqui and R. Lal, "Mechanisms of carbon sequestration in soil aggregates," Critical Reviews in Plant Sciences, vol. 23, no. 6, pp. 481–504, 2004.
36. J. R. Miller, L. W. Morton, D. M. Engle, D. M. Debinski, and R. N. Harr, "Nature reserves as catalysts for landscape change," Frontiers in Ecology and the Environment, vol. 10, no. 3, pp. 144–152, 2012
37. D. McGranahan, D. Engle, S. Fuhlendorf, J. Miller, and D.

Debinski, "Multivariate analysis of rangeland vegetation and soil organic carbon describes degradation, informs restoration and conservation," Land, vol. 2, no. 3, pp. 328–350, 2013.

38. D. A. Mcgranahan, D. M. Engle, S. D. Fuhlendorf, S. J. Winter, J. R. Miller, and D. M. Debinski, "Spatial heterogeneity across five rangelands managed with pyric-herbivory," Journal of Applied Ecology, vol. 49, no. 4, pp. 903–910, 2012
39. R. A. Moranz, D. M. Debinski, D. A. McGranahan, D. M. Engle, and J. R. Miller, "Untangling the effects of fire, grazing, and land-use legacies on grassland butterfly communities," Biodiversity and Conservation, vol. 21, no. 11, pp. 2719–2746, 2012.
40. D. A. McGranahan, D. M. Engle, B. J. Wilsey, S. D. Fuhlendorf, J. R. Miller, and D. M. Debinski, "Grazing and an invasive grass confound spatial pattern of exotic and native grassland plant species richness," Basic and Applied Ecology, vol. 13, no. 8, pp. 654–662, 2012
41. USDA-NRCS, "Web Soil Survey data for Ringgold County, Iowa. Natural Resource Conservation Service, United States Department of Agriculture," 2010,http://websoilsurvey.nrcs.usda.gov.
42. T. J. Stohlgren, K. A. Bull, and Y. Otsuki, "Comparison of rangeland vegetation sampling techniques in the Central Grasslands," Journal of Range Management, vol. 51, no. 2, pp. 164–172, 1998
43. R. Daubenmire, "A canopy-coverage method of vegetational analysis," Northwest Science, vol. 33, no. 1, pp. 43–64, 1959.
44. W. K. Lauenroth and W. C. Whitman, "A rapid method for washing roots," Journal of Range Management, pp. 308–309, 1971.
45. D. Bates, M. Maechler, B. Bolker, and S. Walker, "lme4: Linear mixed-effects models using Eigen and S4," 2013, http://cran.r-project.org/web/packages/lme4/index.html.
46. R Development Core Team, R: A Language and Environment for Statistical Computing, R Foundation for Statistical Computing, Vienna, Austria, 2013.
47. S. Nakagawa and I. C. Cuthill, "Effect size, confidence interval and statistical significance: a practical guide for biologists," Biological Reviews, vol. 82, no. 4, pp. 591–605, 2007
48. S. Nakagawa and H. Schielzeth, "A general and simple method for obtaining R2from generalized linear mixed-effects models,"

Methods in Ecology and Evolution, vol. 4, no. 2, pp. 133–142, 2013

49. J. Oksanen, F. G. Blanchet, R. Kindt et al., "vegan: Community ecology package," 2013, http://cran.r-project.org/web/packages/vegan/index.html.
50. R. B. Jackson, J. Canadell, J. R. Ehleringer, H. A. Mooney, O. E. Sala, and E. D. Schulze, "A global analysis of root distributions for terrestrial biomes," Oecologia, vol. 108, no. 3, pp. 389–411, 1996
51. P. Savadogo, L. Sawadogo, and D. Tiveau, "Effects of grazing intensity and prescribed fire on soil physical and hydrological properties and pasture yield in the savanna woodlands of Burkina Faso," Agriculture, Ecosystems and Environment, vol. 118, no. 1–4, pp. 80–92, 2007
52. D. Foster, F. Swanson, J. Aber et al., "The importance of land-use legacies to ecology and conservation," BioScience, vol. 53, no. 1, pp. 77–88, 2003.
53. K. W. Davies, T. J. Svejcar, and J. D. Bates, "Interaction of historical and nonhistorical disturbances maintains native plant communities," Ecological Applications, vol. 19, no. 6, pp. 1536–1545, 2009
54. D. A. McGranahan, D. M. Engle, S. D. Fuhlendorf, J. R. Miller, and D. M. Debinski, "An invasive cool-season grass complicates prescribed fire management in a native warm-season grassland," Natural Areas Journal, vol. 32, no. 2, pp. 208–214, 2012.
55. S. Trumbore, "Age of soil organic matter and soil respiration: radiocarbon constraints on belowground C dynamics," Ecological Applications, vol. 10, no. 2, pp. 399–411, 2000
56. C. T. Garten Jr. and S. D. Wullschleger, "Soil carbon inventories under a bioenergy crop (Switchgrass): measurement limitations," Journal of Environmental Quality, vol. 28, no. 4, pp. 1359–1365, 1999
57. S. T. Pickett, "Space-for-time substitution as an alternative to long-term studies," inLong-Term Studies in Ecology: Approaches and Alternatives, pp. 110–135, Springer, New York, NY, USA, 1989.

Chapter 5

Afforestation for the Provision of Multiple Ecosystem Services: A Ukrainian Case Study

Maria Nijnik[1], Arie Oskam[2], and Anatoliy Nijnik[3]

[1]Social, Economic and Geographical Sciences, the James Hutton Institute, Craigiebuckler, Aberdeen AB15 8QH, Scotland, UK

[2]Agricultural Economics and Rural Policy, Wageningen University, Hollandseweg 1, 6706 KN Wageningen, The Netherlands

[3]Environmental Network Ltd., the Hillocks, Tarland, Aboyne AB34 4TJ, Scotland, UK

ABSTRACT

This paper presents an economic analysis of the planting of trees on marginal lands in Ukraine for timber production, erosion prevention, and climate mitigation. A methodology combining econometric analysis, simulation modelling, and linear programming to analyse the costs and benefits of such afforestation has been adopted. The research reveals that, at discount rates lower than 2%, establishment of new forests is economically justified in the majority of forestry zones. Incorporating the effects of afforestation on mitigating climate change increases social benefits. However, the results indicate that whilst soil protection benefits to agriculture from afforestation in the Steppe are expected to be high, carbon sequestration and timber production activities in the Steppe are cost inefficient due to low rates of tree growth and relatively high opportunity costs of land. The opportunity costs of land are also high in the Polissja where afforestation is cost inefficient at 2% and higher discount rates.

INTRODUCTION

Forestry is multifunctional by nature, but has traditionally been a sector of the economy whose primary objective is to maximize profits from timber production. Today, the focus of forestry is much wider, with evaluation of a broad range of ecosystem services being the mainstream in forestry economics. It is being increasingly recognised that, in addition to timber production, woodlands contribute to global carbon budgeting, provide biodiversity and aesthetic values, and serve as the basis for developing local entrepreneurial opportunities, tourism, recreation, and rural livelihoods [1]. The manifold ecosystem services do not represent a minor issue but rather are an important precondition for diverse human activities which predetermine multifunctionality of forests [2]. The relative societal weight of forestry and its manifold contribution to livelihoods are most clearly reflected in policies focusing on the development of new forests [1, 3]. Afforestation for multiple purposes is reflected in policies of many countries [4] where multifunctional development of forestry is evidenced through institutional analyses and public opinion surveys [5] and is expressed in real terms on the ground [6].

The origin of discussions on "what forest's multifunctionality is" and "how to manage forests for multiple purposes" goes back to the 1940s. An important question is whether multifunctionality should be considered in the vertical sense, with each lot of land or forest stand fulfilling two or more functions, or whether the term should be used to describe a pattern of diversity in the horizontal sense, in which different areas are dedicated to different functions. Dana implied [7] a vertical interpretation of multifunctionality, while Pearson's view [8] was horizontal: "effective multiple use is merely organized and coordinated specialization." The vertical vision of multifunctionality seemed to be dominant in the last part of the 20th century, although the horizontal interpretation also had its advocates [9]. Bowes and Krutilla [10], for example, argue that forestry is capable of producing successfully a long list of outputs, many of which are complementary in production, whereas Sedjo questions [11] whether there is a role for specialization in multifunctional forest management.

In the current paper, forest multifunctionality is considered in the vertical sense [7]. Characterised in this way, multifunctionality is seen as an economic concept, capturing the multiple processes taking place in forests, and shifting the emphasis of forestry away from maximising production of material (commodity) goods towards its broader objectives, that is, the provision of environmental services (erosion alleviation and climate change mitigation). Climate change, changing socioeconomic conditions, and fragile ecosystem stability have resulted in a need for a more enlarged and continuous forest cover in countries like Ukraine [12] and in new, adaptive approaches to forest management [13, 14]. A suite of forest management alternatives can be ranging from intensive even-aged management to close-to-nature forestry, and nonintervention systems [15]. The selection of an appropriate management system is considered to be fundamental to implementing multifunctional forestry [16]. Multifunctionality implies that the objectives of numerous forestry stakeholders are incorporated into an adaptive and participatory planning process so that ecosystem services are enhanced across the landscapes [14, 17].

Decision Support Systems (DSSs) are suitable platforms for the integration of information, models, and methods required to support complex forest management solutions including those of afforestation [18, 19]. With its focus on North-Western Europe, the AFFOREST DSS seeks to answer on where, how, and how much to afforest in order to maximize carbon sequestration and groundwater recharge and to minimize nitrate leaching [20]. ENCOFOR gives support on where and how to afforest land in developing countries in order to combine successful greenhouse gas mitigation with poverty alleviation and environmental restoration [21]. The SimForTree DSS is designed to compare different forest management and policy scenarios at different spatial scales under changing environmental conditions, particularly climatic [22]. Large-scale scenario models incorporating economic considerations of forest creation and management are becoming available [23].

Afforestation has been widely analysed in the literature, and it is commonly considered by contemporary forestry science among the most straightforward policy measures, including those addressing climate change [24]. It is relatively cheap (cost efficient), clean (it may concurrently provide other ecosystem services), proven (many countries have the legacy of tree growing), effective in the short-term (providing almost immediate effect after the tree planting) and is a rather low labour and energy consuming strategy [25]. It can be incorporated in multifunctional forest use to simultaneously enlarge timber production and bring a variety of other benefits [26].

The purpose of afforestation entails the choice of species to be planted, as does the location, tree-growing conditions and other factors. Carbon sequestration positively correlates with the growth rates of trees; it is therefore advocated to plant the most fast-growing tree species where appropriate or to establish short rotation plantations for the purpose of climate change mitigation alone [27, 28]. However, there is sufficient evidence that forests significantly reduce soil erosion and prevent soil run-off after heavy rains [12, 29]. Tree plantations are still the main providers of materials for construction and industries, and of fuel for economies and households [30]. Forests provide multiple ecosystem services, and in consideration of a more holistic approach to afforestation, a number of recent studies have a wider focus by addressing several forest functions in one instance.

Michetti and Nunes [31] analyse the role of afforestation and timber management activities in stabilizing climate. Lindner et al. [32] cover the capacity of forests to provide economic, social, and ecological services (wood production; nonwood forest products; carbon sequestration; biodiversity; recreation; protective functions) under the impacts of climatic change. The European Forestry Institute used the European Forest Information (EFISCEN) modelling approach [33] to gain insights into the effects of changing management practices in forests across 30 countries under different scenarios [34]. The need to maintain the national diversities that constitute European forestry within harmonised strategies is being recognised.

Ukrainian scientists contributed to the scenario analysis through the development of long-term predictions for their forests. Their scenarios are consistent with the projections developed at the International Institute for Applied Systems Analysis (IIASA), the Ukrainian National Agricultural University, and the Ukrainian National University of Forestry [35–37]. Numerous forestry studies support the idea that a larger forest area would be attractive in Ukraine [36, 38], but limited knowledge is available about the economics of creating multifunctional forests to address wider sustainability objectives.

Thus, based upon both international and national studies on multipurpose afforestation, the current paper considers the economics of tree planting in Ukraine for the provision of timber production, soil protection and carbon sequestration services. Firstly, "user" benefits of afforestation which accrue to two primary sectors of rural economy, that is, forestry and agriculture, are considered, with evaluation of the role of forests in soil protection complementing an assessment of the timber supply benefits accruing from establishment of forest plantations. The paper then defines key objectives of woodlands expansion and analyses the potential for tree planting, estimating its costs and benefits. This is followed by a pilot cost-benefit analysis (CBA) of the establishment of forest plantations: if afforestation adds to the welfare of society, it is economically sound. This is observed in some areas in Ukraine, and we discuss where and why this is the case. Finally, by considering carbon sequestration in trees under the storage policy scenario, the paper goes beyond the "user" benefits of afforestation that accrue to the forestry and agriculture sectors, linking the planting of trees in Ukraine to global climate change mitigation policy objectives.

BACKGROUND

The history of establishment of forest plantations in Ukraine dates back to the 17th century; however, until the 1920s, tree-planting activities were isolated and episodic [39]. The protection of land through woodland creation was considerably hampered by economic and social conditions and then by the shift in forest policy from afforestation towards natural forest regeneration. In the 1990s, the decreasing scale of afforestation was also caused by difficulties of the transition process from a command-and-control towards a market economy, such as the lack of well-defined and ensured property rights; shortage of investment and economic incentives; increasing attention paid by forest management to short-term financial objectives [40].

Nevertheless, the reforestation coefficient remains at 94%; planting on forestland amounts to about 28.5 kha; protective plantings on eroded and unproductive agricultural lands total 7 kha; shelterbelt plantings total 1.1 kha per year [36]. The necessity of afforestation is stressed in the President's Decree [41], which aimed to reform forestry in Ukraine, and in the State Programme "Forests of Ukraine, 2002–2015" [42]. Forestry legislation rests on the principle of sustainable forest management, and multiple forestry objectives are recognised by law [43].

Ukraine has good forest growing conditions and productive forests [44]. The territory was extensively covered with forests a millennium ago. Today, forest cover comprises 16.5%, and this is among the lowest estimates in Europe [45]. Given that 15% of Ukraine's territory is under extreme anthropogenic pressures, and considering the role of forests for environmental quality, the development of woodlands is considered important [46]. Ukraine has a level of cultivation of 54.8% by area and faces partial erosion on 35% of its arable land. Annually, 4 Mt of fertile soil are washed away. The damage to agriculture from erosion exceeds 8 M€ [39]. The intensity of soil erosion varies across the territory for which the annual increment of eroded land is 90 kha. According to the National Academy [47] wooded cover should increase to 20%, since this would alleviate spatial spreading of erosion and its intensity.

In recent years, afforestation in Ukraine has played an important role also as a carbon dioxide reduction measure [48]. Since 2001, afforestation has become an eligible component of climate policy, meaning that over and above emissions reduction, an enhancement of GHG "sinks" and "reservoirs" via tree planting is becoming considered as increasingly important. Also, as a consequence of its afforestation and sustainable forestry strategy, Ukraine could become self-sufficient in wood and become a price setter in the European wood production. This is because the demand for wood in Europe is rising by 0.8–2.6% per year, with Europe likely to remain a high wood cost region. The potential niche for Ukraine's forestry in an international perspective is therefore its low cost of delivered wood [49].

METHODOLOGY AND THE RESULTS

A methodology combining econometric analysis, simulation modeling, and linear programming for a cost-benefit analysis of afforestation has been adopted, as described in the following sections, which also present the results of the research. The modelling approach is straightforward, practical, and policy relevant. It could be easily modified, adjusted, and applied to other cases and in other countries to review evidence and the opportunities available with respect to the delivery of multiple services from land use policy decisions by assessing existing forest management options and identifying new ones.

Afforestation Potential

Afforestation potential was assessed across Ukraine's forestry zones. We developed comprehensive forestry zoning in an earlier, separate study (Figure 1) to provide a background for consideration of policy measures to enhance sustainability of forest use [39].

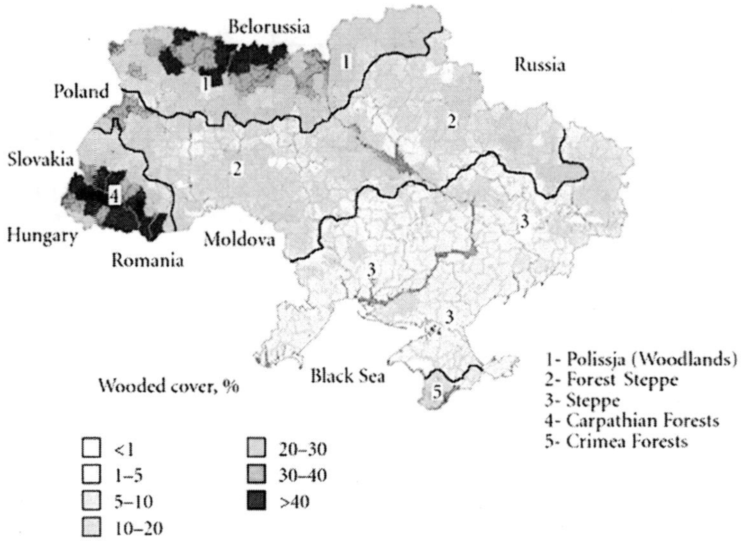

Figure 1: Forestry zoning of Ukraine Source: adapted from [39] and the records of the National Academy of Sciences [47].

This spatial classification divides the territory of Ukraine into the following main forestry areas: the Polissja (Woodland), Forest Steppe, Steppe, Crimea, and the Carpathians, and further divides these main forestry areas into spatial units of lower levels of hierarchy using various factors of land use and forestry development. The land suitable for tree planting was deemed to include bare land of the "State Forest Fund" (SFF, that is, public forest estate) that is under the management of the State Committee of Forests. The land deemed suitable for afforestation also included bare and marginal agricultural land currently used for forage and pasture and some land used for wheat production for which net returns are low (Table 1).

Table 1: Potential for afforestation by land use across zones, kha

Zones	State forest fund			Agricultural land			Total
	Ravines	Sand	Rocky	Eroded	Deflated	Rocky	
Polissja (Woodland)	65.0	82.0	0.5	73.7	0.7	26.1	248.0
Forest Steppe	95.0	84.0	0.6	451.6	18.3	61.0	710.5
Steppe	24.0	64.0	n.a.	669.4	40.6	137.5	935.5
Carpathians	1.6	n.a.	1.4	24.6	n.a.	143.4	171.0
Crimea	0.8	n.a.	1.8	13.1	1.8	206.8	224.3
Ukraine (totally)	186.4	230.0	4.3	1232.4	61.4	574.8	2289.3

Source. Computed on the basis of data from the National Academy of Sciences of Ukraine [47]

The table shows that a total area of 2.29 Mha was initially identified as suitable for tree planting. After estimating the net present value (NPV) of afforestation, this area was reduced, to account for land in which the opportunity costs appear to be comparatively high (see below for further details). Given tree growing conditions and assumptions based on interviews with forest specialists [44], pine was considered appropriate for planting in the Steppe, Crimea and Polissja; pine and oak in the Forest Steppe; beech, fir, and/or spruce in the Carpathians.

Costs of Afforestation

The costs of afforestation of marginal and bare land in the SFF include tree-planting costs (€100–200/ha) and silvicultural expenses (€ 12.5–30/ha annum). These costs vary; but given that within each zone conditions are relatively stable, costs are assumed to be the same within each zone. Marginal agricultural land has alternative options to afforestation, therefore, in addition, net returns associated with current use of land were considered. Net annual returns for current wheat production were computed on the basis of land productivity data, costs of wheat production, and output prices. Estimation of costs for land used for forage and pasture was based on land productivity, and prices which agricultural enterprises pay for the equivalent cattle feed (Table 2).

Table 2: Net annual returns to current agricultural activities, €/ha

Forestry zone		**Forage and pasture**	**Wheat**
Polissja (Woodland)	Eroded and deflated land	8.0	37.8
Forest Steppe	Rocky land	7.8	n.a.
	Eroded land	10.0	52.1
	Deflated lands	9.2	14.7
Steppe	Rocky land	8.0	n.a.
	Eroded land	20.0	61.5
	Deflated land	6.0	27.2
Carpathians	Rocky land	n.a.	n.a.
	Eroded land	7.8	0
Crimea	Rocky land	7.0	0
	Eroded and rocky	7.0	0

Source. Based on the data from the National Academy of Sciences of Ukraine [28]

Computation was in Ukraine's currency Hryvnya (which in 2007 corresponded to 0.14 €). The present value (PV) costs occurring over 100 years of stipulated ages of timber harvesting were estimated at several discount rates based on [36], as seen in Table 3.

Table 3: Afforestation costs, M €

Forestry zone	Annual costs by zone			PV costs			
	Opportunity	Planting	**Care and protection**	**r =0%**	**r =2%**	**r =4%**	**r =6%**
Polissja (Wooland)	1.4	16.1	2.0	356.3	162.7	99.5	72.7
Forest Steppe	6.4	32.8	4.1	1084.3	486.0	290.5	207.5
Steppe	14.1	49.8	7.1	2173.3	965.0	570.2	402.7
Carpathians	0.8	7.5	0.9	177.9	80.9	49.2	43.8
Crimea	0.8	19.6	2.5	345.0	159.9	99.4	73.7
Ukraine	23.5	125.8	16.6	4136.8	1854.5	1108.8	792.4

Source: Computed on the basis of data from the National Academy of Sciences of Ukraine [28]

The results show that at a 4% discount rate, the PV of afforestation costs is €484 per ha on average for the country. The highest PV of costs is in the Steppe at €609.5 per ha, with the lowest, at €288 per ha, in the Carpathians. The divergence in cost estimates across zones is explained by the diversity of tree-growing and socioeconomic conditions.

Timber Supply Benefits

A sum of monetary value from additional timber yield and monetary estimates of soil protection pertaining to arable land comprise the total benefits of afforestation (when only "user benefits" to forestry and agriculture are considered). For the monetary value of timber yield changes, the model multiplies estimates of a physical crop change based on acreage in production by the price of timber [50]. This implies that timber use and prices are constant. Allowing in the long run for a stable average annual timber cut of 2 m^3/ha (ca. 50% of mean annual increment, MAI), about 4.6 Mm^3 of additional timber could be produced, bringing annual returns of 23 M€, if the stumpage value of timber is ca. €5/m^3 [49].

Benefits were also computed across zones over a 100-year period. Table 4 shows the results when only commercial timber cut is taken into account. The following assumptions were made: stand composition in the Forest Steppe comprises 50% pine and 50% oak trees; half of the Steppe is planted with trees which would be harvested; beech stands in the Carpathians are planted on 50% of the area, as are fir stands. The Crimea, where forests play primarily a role in enhancing environmental quality, timber harvesting was not considered.

Table 4: Estimates of the returns from timber harvesting

Zone	Species	Stock of stands in 100 years,	Returns in the year of harvesting	PV returns	PV returns by zone			
		m^3/ha	€ /ha	€ /ha,	M €			
				4%	0%	2%	4%	6%
Polissja	Pine	250	1250	24.75	310.0	42.8	6.1	0.9
Forest Steppe	pine	350	1750	34.65	612.9	84.6	12.1	1.8
	Oak	350	1750	34.65	612.9	84.6	12.1	1.8
Steppe	pine	250	1250	24.75	584.7	27.7	11.6	1.7
Carpa-thians	beech	350	1575	31.18	134.7	18.6	2.7	0.4
	fir	400	2000	39.6	171.0	23.6	3.4	0.5
Ukraine					2304.0	318.0	45.6	6.8

The table suggests that, at the discount rate of 0%, PV returns from timber harvesting are 23.04 M€ (comparable with the annual returns of 23 M€ estimated above). The highest returns per acreage are in the Forest Steppe and the Carpathians. However, timber benefits alone do not justify tree planting in any of the zones.

Soil Protection Benefits

The notion that the scale of erosion depends on forest cover [47] was put to an empirical test by using a semilogarithmic regression (Figure 2). The robustness of the functional form adopted was justified, and economic attractiveness of planting trees to mitigate erosion was then assessed.

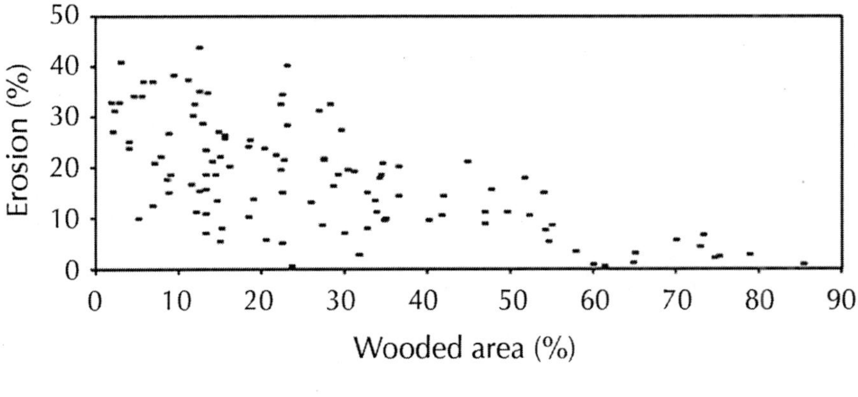

Figure 2: Relationship: wooded cover—erosion.

The results of the estimation show a statistically significant (at 1% significance level) negative relationship between the share of eroded land (E, %) and the share of wooded land (W, %):

$$\log(E) = 3.4653 - 0.0329 * W; \quad \text{or } E = 31.986 e^{-0.0329W},$$
$$(29.13) \quad (-9.38) \qquad R^2 = 0.45 \tag{1}$$

The t-statistic of −9.38 suggests that the negative coefficient on W is significantly different from 0, and with the increase of forest cover, the erosion rates decrease. The value of R^2 indicates that more factors influence erosion rates and there is room for improvement of the model.

For the Carpathian Mountains, where the conditions differ substantially from elsewhere in the country (lowland), the estimated equation is as follows:

$$\log(E) = 4.3702 - 0.0523 * W; \text{ or } E = 79.059 e^{-0.0523W},$$

$$(5.46) \quad (-3.99) \qquad R^2 = 0.5 \qquad (2)$$

Simulated rates of erosion at different levels of wooded cover are shown in Table 5.

Table 5: Simulated rates of soil erosion

Wooded area (W), %	Erosion (E), Ukraine, %	Erosion (E), Carpathians, %	Elasticity, Ukraine, %	Elasticity, Carpathians, %
0	32.0	79.1	-1.05	-4.13
5	27.1	60.9	-0.89	-3.18
10	23.0	46.9	-0.76	-2.45
15	19.5	36.1	-0.64	-1.89
20	16.6	27.8	-0.54	-1.45
25	14.1	21.4	-0.46	-1.12
30	11.9	16.5	-0.39	-0.86
35	10.1	12.7	-0.33	-0.66
40	8.6	9.8	-0.28	-0.51
45	7.3	7.5	-0.23	-0.39
50	6.2	5.8	-0.20	-0.30
55	5.2	4.4	-0.17	-0.23
60	4.4	3.4	-0.15	-0.18
65	3.8	2.6	-0.12	-0.14
70	3.2	2.0	-0.11	-0.11
75	2.7	1.6	-0.09	-0.08
80	2.3	1.2	-0.08	-0.06
85	2.0	0.9	-0.06	-0.05
90	1.7	0.7	-0.05	-0.04
95	1.4	0.5	-0.05	-0.03
100	1.2	0.4	-0.04	-0.02

From the estimated equations, marginal changes in erosion rates relative to marginal changes in wooded cover rates are for Ukraine: dE/dW=-0.0329E, and for the Carpathians: dW/eW=-0.0523E. These estimations show the "elasticity" of erosion with respect to wooded cover. These results indicate that until wooded cover reaches 27% in the Carpathians and ca. 2% in Ukraine, the erosion is elastic, that is, when wooded cover is increasing marginally, the erosion is reduced proportionally as much. This is observed up to the point when the share of eroded land is around 30% in Ukraine, and as far as it falls below 19% in the Carpathians. The results suggest that if there were no woods in rural areas the share of eroded lands would comprise 79% in the Carpathians and 32% on average in Ukraine.

By using the results of the regression analysis, indicative estimates of the soil protection role of forests were computed. In the Polissja where wooded cover comprises ca. 26%, the "elasticity" of erosion is −0.43%. This means that a 1% increase in wooded cover leads to a 0.43% decrease in erosion rates. A 1% increase of forest cover, that is, an increase of 0.029 Mha, will mitigate erosion on 0.2 Mha of land. The net annual returns from that land were then calculated on the basis of data from Table 2 and were considered as measures of soil protection benefits to agriculture from marginal expansion of forests in the Polissja. The corresponding estimations were made for other zones, and the equation is:

$$X = \varepsilon . \frac{E}{W}, \qquad (3)$$

where X is the indicative measure of soil protection benefits to agriculture from marginal expansion of forest cover; ε "elasticity" of erosion with respect to forest cover, that is, 1% increase in W leads to ε % decrease in E, % (Table 5); W share of wooded land, %; E share of eroded agricultural land, %.

Forests start providing protection benefits after the age of 5 years, and with their gradual regeneration, forests keep providing these benefits indefinitely [12]. These considerations were taken into account when computing economic estimates of the soil protection function of forestry for agriculture. The potential for forest expansion was taken from Table 1, and the assumption was made that, in no mountainous areas, 30% of agricultural land is used for wheat production. Given the assumptions presented above, the soil protection benefits of afforestation for agriculture are shown in Table 6 and are the highest in the Steppe.

Table 6: Estimates of soil protection benefits to agriculture

Forestry zone	Annual average benefits, € /ha		Annual benefits
	Wheat	Forage/pasture	M € /zone
Polissja (Woodland)	7.6	1.6	0.8
Forest Steppe	33.0	9.0	11.5
Steppe	58.2	17.0	27.5
Carpathians	0	9.7	1.7
Crimea	0	12.2	2.7
Ukraine			44.2

Economic Analysis of Afforestation for Timber Production and Erosion Alleviation

The economic analysis of afforestation was carried out at various hierarchical levels. First, "user" net present value (NPV) benefits of afforestation to forestry and agriculture in Ukraine were considered. The general idea of the LP model discussed by us in detail in [44] was to consider the establishment of future forests on bare and marginal agricultural lands in Ukraine, in such a way that allows the attainment of maximum cumulative NPV of benefits from new plantations over the period of 100 years, subject to constraints. The land available for conversion into forest and tree species most suitable for growing in the particular conditions, identified earlier in the current paper, were considered, together with the following forest management regimes. The first forest management regime was a basic silviculture that is based on quick replanting of trees, after timber harvesting. The second regime was that of planting trees and the attendant silvicultural operations recommended by Ukraine forest legislation. The third regime was that of a basic silviculture, but with the timber rotation period corresponding to maximum sustainable yield.

Major constraints of this model are acreage. The model implies, for instance, that in the Carpathian Mountains, beech forests do not grow at high altitudes or that there is no land suitable for wheat production in the mountains.

The results of LP modelling provide evidence that under the assumptions and at a discount rate of 4%, it is economically sound to establish monoculture plantations on the perceived bare land in the Forest Steppe (0.28 Mha), Steppe (0.13 Mha), and the Carpathians (0.01 Mha). The shadow price of bare land (245.2 €/ha) is highest in the Steppe. Establishment of monoculture plantations appeared to be a more economically sensible solution, with basic silviculture proven to perform better economically in all forestry zones. These results can be explained by the fact that the scope of this phase of the research was limited to user values that accrue to Ukraine's forestry and agriculture. The purpose was to form a basis for economic appraisal of practical aspects of land use management decisions. The analysis has been further extended, and in the following section, the research is presented for a higher hierarchical scale, so that in addition to "user" benefits of afforestation, carbon sequestration benefits in trees under a storage policy scenario are incorporated in the analysis.

Comparing Costs and Benefits of Afforestation

In addition to domestic gains to forestry and agriculture, afforestation provides climate change mitigation benefits. The economics of carbon sequestration forestry scenarios for Ukraine as a stand-alone analysis is considered in Nijnik [51]. In the current paper, economic evaluation of tree planting for multiple purposes, including for carbon sequestration under the storage policy scenario, is presented. The storage policy scenario presumes a one-time planting of trees for a period of 100 years, without accounting for future use of wood and land [52]. The analysis was carried out across forestry zones with maximising of NPV of afforestation as the criterion. It determines the present value of net benefits by discounting the stream of benefits (B) and costs (C) back to the beginning of the base year t=0:

$$NPV = \sum_{t=0}^{n} B_t / (1+r)^t - \sum_{t=0}^{n} C_t / (1+r)^t \qquad (4)$$

Benefits of afforestation are expected to accrue over a long period of time, and a period of 100 years was chosen to capture most of these benefits and costs. In addition to timber production and soil protection benefits from the potential forests to alleviate soil erosion discussed earlier, the carbon (C) uptake benefits from afforestation have been approximated using the following procedure.

The functional forms for stand growth of tree species were estimated, using the equations provided by the authors in Nijnik [53]. The coefficients of Lakida et al. [54] were used to translate the stem biomass into total above-ground biomass. The volume of stem wood was multiplied by 0.2 tC/m^3 for its translation into carbon [55]. Carbon sequestered by the root was estimated, depending on tree species, either on the basis of the relationships presented in Van Kooten and Bulte [52] or in Lakida et al. [54]. Then, based on Nijnik [44, 51], the sequestered carbon was computed across zones. The price of 15 € per tonne of carbon was assumed to be stable and was used to calculate carbon uptake benefits based on Sandor and Skees [56]. The fact that an increase in the amount of carbon credits available to buy, especially pertaining to "hot air," is pushing the price of carbon credits down was not taken into account. The discount settings of 0%, 2%, and 4% were applied when calculating the carbon storage option. The PV benefits from afforestation are shown in Table 7.

Table 7: Afforestation benefits, PV M €

Forestry zone	r%	Production	Soil protection	Carbon uptake
Polissja (Woodland)	0	310	84	362.6
	2	42.8	36.2	49.1
	4	6.1	20.6	6.6
Forest Steppe	0	1125.8	1150	1255.9
	2	169.2	495.6	170.1
	4	24.2	281.8	23.0
Steppe	0	584.7	2750	1237.4
	2	80.7	1185.2	167.5
	4	11.6	673.9	22.7
Carpathians	0	305.7	170	660.7
	2	42.2	73.3	89.4
	4	6.1	41.7	12.2
Crimea	0	0	270	437.3
	2	0	116.4	59.2
	4	0	66.2	8.0

The results presented in Table 7 suggest that, for example, in the Polissja the highest benefits would accrue from the increased timber production and carbon uptake, whilst in the Steppe the highest benefits would accrue from the forest function of soil protection.

The results vary substantially across the territory of Ukraine and depend upon the applied discount rates. The NPV of afforestation is positive in the majority of zones for discount rates of up to 2%. At these discount rates, creation of multifunctional forest plantations would enlarge social benefits, including those of climate change mitigation, and would add to the welfare of society. At a discount rate of 4%, the area of forest plantations would be 1.82 Mha (excluding the Crimea and the Polissja). In the Carpathian and Crimean Mountains, commercial timber harvesting is restricted, and economic benefits from timber are therefore modest. Agricultural production is also limited in the mountainous regions. Consequently, the benefits that accrue to agriculture from the forest function of soil protection are moderate (Table 8).

Table 8: Economic evaluation of afforestation across zones, PV M €

Forestry zone	r%	Total benefits	Costs	NPV
Polissja (Woodland)	0	756.6	356.3	400.3
	2	128.1	162.7	−34.6
	4	33.3	99.5	−66.2
Forest Steppe	0	3531.7	1084.3	2447.4
	2	834.7	486	348.7
	4	329.0	290.5	38.5
Steppe	0	4572.1	2173.3	2398.8
	2	1433.4	965	468.4
	4	696.6	570.2	126.4
Carpathians	0	1136.4	177.9	958.5
	2	204.9	80.9	124
	4	59.9	49.2	10.7
Crimea	0	707.3	345	362.3
	2	175.6	159.9	15.7
	4	74.2	99.4	−25.2

CONCLUSIONS

Afforestation in Ukraine, where vast areas are currently suitable for tree planting, is seen as a means to contribute to sustainable land management and climate change mitigation. The afforestation costs are fairly low, apparently due to good forest growing conditions and relatively low labour costs. An expansion of forest cover is important for soil protection. Annually, 1 ha of forest in Ukraine provides soil protection benefits to agriculture of €1.6 to €58.2. Such a broad range can be explained by the variety of conditions. A low share of forest cover might be among the causes of erosion, and planting trees is a possible measure to alleviate this, particularly in the Steppe. When only timber production gains and gains from the protection of agricultural land against erosion are taken into account, at 2% through 4% discount rates, the benefits from afforestation are high in the Steppe, Forest Steppe, and the Carpathians, where the tree planting is economically justified on ca. 1.82 Mha of land. When a discount rate of 4% is used, economic justification of tree planting would be limited to bare land in these zones, with a total area of 0.42 Mha.

The results are more positive when afforestation considerations include the rewards for climate change mitigation. But with or without consideration of carbon uptake, at discount rates lower than 2%, afforestation costs are covered by the returns in the majority of regions. The results indicate that whilst soil protection benefits from afforestation in the Steppe are expected to be high, carbon sequestration and timber production activities are not cost efficient due to low rates of tree growth and relatively high opportunity costs of land. The opportunity costs of land are also high in the Polissja where afforestation is cost-inefficient at 2% and higher discount rates.

DISCUSSION

In Ukraine, and many other former command-and-control economies (to which this study would convey), large-scale agriculture under the previous regime supported the conversion of forest or grassland to agricultural land. Currently, the decreasing agriculture [45] will likely cause the increase of abandoned land. An expansion of woodlands and, consequently, a rising role of forestry could therefore be predicted. International regulations and national programmes supporting the conversion of agricultural land back to forest focus largely on remote areas. Therefore, afforestation projects need to be incorporated in regional schemes of sustainable rural development, where socioeconomic, environmental, and climate change related components of land use changes are to be considered jointly.

In order to be viable afforestation projects need to be coherent, effective, cost efficient, widely acceptable by the public and consistent with other aspects of sustainable development policy [28]. Numerous examples indicate that climate policy measures will likely be accepted by the public and will consequently be implemented, if they are consistent with the programmes that focus on issues, other than climatic stresses [57]. Many scholars support this view, emphasising the "win-win" opportunities of forest carbon projects, which may all at once provide biodiversity conservation and rural development benefits [58]. Policy measures then should aim the "win-win" situations, which would benefit rural development, the environment, people, and the economy all together.

Afforestation for multiple purposes could enlarge total benefits and prevent potential conflicts related to the trade-offs between biodiversity and carbon sequestration benefits or between landscape amenity values and those of carbon uptake. Although a multipurpose afforestation may result in lower rates of carbon sequestration, it is expected to be more attractive to people, because in the majority of cases it will provide additional benefits and promote sustainable rural development. A multi-purpose afforestation is often seen as a sustainable way of land reclamation and of increasing of the productivity of abandoned land [59, 60], whilst utilization of biomass from new plantations can provide employment opportunities and create new options for land development, being also a sustainable alternative for nuclear energy.

The potential for afforestation in Ukraine appeared to be significant. This allows us to argue that forestry activities can contribute to climate change mitigation and can provide additional social and environmental benefits by contributing to economic development in remote regions, which are most strongly affected by the transition processes. However, new insights are needed into the connection between climate policies and strategies to promote sustainable forestry and sustainable land use and sustainable rural development. Efficient and feasible forestry-based initiatives and inter-sectoral cooperation need to be well embedded into existing policy areas, and if flexible mechanisms are implemented, a considerable scope exists for multifunctional land use systems and "win-win" solutions.

In transition countries affected by regional socioeconomic disparities, the economic and social issues of forestry development are particularly important in view of accumulating financial assets and providing social opportunities for sustainable policy and promarket reforms. Afforestation seeks to enlarge economic gains to forestry through an enlarged wood production and to the agricultural sector, because of soil protection and hydrological forest functions. It also deemed to be beneficial in a broader sense, for example, concerning the mitigation of climate change.

The Kyoto Protocol provides opportunities for countries to cope with the changing climate from an economic perspective. Its flexible mechanisms were designed to help Annex B countries, including Ukraine, to meet their emissions reduction targets at least cost [61]. It allows (non-EU) countries like Ukraine to sell carbon offsets to industrialised countries and, therefore, raise funds needed for its forestry sector and a wider economy. Carbon sequestration through afforestation in Ukraine could be beneficial also for Annex B countries in view of stabilising their collective emissions in the cheapest possible way via the trading of carbon-offset services. European investors are showing their interest to invest in afforestation projects in Ukraine. However, the potential gains from international projects are not seen as priorities for climate policies in either of these countries [62, 63]. Therefore, unless the necessary institutional infrastructure is developed and the barriers for investment are identified and addressed, Ukraine cannot expect to benefit widely from carbon crediting.

Thus, there is a difference between the benefits of afforestation identified in this paper and the benefits for those planting the trees. Although tree planting would enlarge social gains, welfare maximisation conditions would hardly be guaranteed because of market failures. The problem: "who pays for tree planting and who receives the benefits" can scarcely be solved in Ukraine through the market and would therefore need to be regulated by government [37]. It seems that government would have to justify public policy concerning tree planting and balance costs and benefits to provide incentives for those planting the trees. Also, afforestation would have to be elaborated in close dialogue with stakeholders and with public involvement including in-depth consideration of various scenarios for woodlands expansion. Such deliberative processes will increase the capabilities of policy actors, assisting them in the delivery of sustainability objectives to forest management practices on the ground.

REFERENCES

1. Millennium Ecosystem Assessment (MEA), Millennium Ecosystem Assessment Synthesis Report, 2005,http://maweb.org/documents/document.356.aspx.pdf.
2. C. Folke, J. Colding, and F. Berkes, "Synthesis: building resilience and adaptive capacity in social-ecological systems," in Navigating Social-Ecological Systems: Building Resilience for Complexity and Change, F. Berkes, J. Colding, and C. Folke, Eds., pp. 352–387, Cambridge University Press, 2003.
3. European Commission (EC), Environmental Policy Review: Communication from the Commission to the European Parliament COM, vol. 195. Brussels, Belgium, 2007,http://www.nhlbi.nih.gov/meetings/workshops/cardiorenal-hf-hd.htm.
4. M. Kaljonen, E. Primmer, G. De Blust, M. Nijnik, and M. Kulvik, "Multifunctionality and biodiversity conservation—institutional challenges," in Nature Conservation Management: From Idea to Practical Issues, T. Chmelievski, Ed., pp. 53–69, Helsinki-Aarhus, Lublin, Poland, 2007.
5. M. Nijnik, L. Zahvoyska, A. Nijnik, and A. Ode, "Public evaluation of landscape content and change,"Land Use Policy, vol. 26, no. 1, pp. 77–86, 2008.
6. A. Mather, G. Hill, and M. Nijnik, "Post-productivism and rural land use: cul de sac or challenge for theorization?" Journal of Rural Studies, vol. 22, no. 4, pp. 441–455, 2006.
7. S. T. Dana, "Multiple use, biology and economics," Journal of Forestry, vol. 41, pp. 625–627, 1943.
8. G. A. Pearson, "Multiple use in forestry," Journal of Forestry, vol. 42, pp. 243–249, 1944.

9. J. R. Vincent and C. S. Binkley, "Efficient multiple-use forestry may require land-use specialization,"Land Economics, vol. 69, no. 4, pp. 370–376, 1993.
10. M. D. Bowes and J. V. Krutilla, The Multiple-Use Economics of Public Forest lands: Resources for the Future, Washington, DC, USA, 1989.
11. R. Sedjo, "Transgenic trees and trade problems on the horizon," Resources, vol. 155, pp. 9–13, 2004.
12. S. Gensiruk and S. Ivanytsky, Forest Management and Setting up a Proper Share of Forest Cover, Lviv Publisher, L'viv, Ukraine, 1999.
13. European Environmental Agency (EEA), "Europe's ecological backbone: recognising the true value of our mountains," EEA Report 6/2010, p. 248, Copenhagen, Denmark, 2010.
14. L. Fazey, J. Gamarra, J. Fischer, M. Reed, L. Stringer, and M. Christie, "Adaptation strategies for reducing vulnerability to future environmental change," Frontiers in Ecology and the Environment, vol. 8, no. 8, pp. 414–422, 2010.
15. J. Carnus, J. Parrotta, E. Brockerhoff et al., "Planted forests and biodiversity," Journal of Forestry, vol. 104, no. 2, pp. 65–77, 2006.
16. G. Buttoud, "Multipurpose management of mountain forests: which approaches?" Forest Policy and Economics, vol. 4, no. 2, pp. 83–87, 2002.
17. G. Weiss, "The political practice of mountain forest restoration—comparing restoration concepts in four European countries," Forest Ecology and Management, vol. 195, no. 1-2, pp. 1–13, 2004.
18. J. G. Borges, A. Falcão, C. Miragaia, P. Marques, and M. Marques, "A decision support system for forest resources management in Portugal," in System Analysis in Forest Resources, G. J. Arthaud and T. M. Barrett, Eds., vol. 7 of Managing Forest Ecosystems, pp. 155–164, Kluwer Academic, 2003.
19. K. M. Reynolds, M. Twery, M. J. Lexer, et al., "Decision support systems in natural resource management," in Handbook on Decision Support Systems, F. Burstein and C. Holsapple, Eds., vol. 2 ofInternational Handbooks on Information Systems Series. Handbook on Decision Support System, pp. 499–534, Springer, 2008.

20. S. Gilliams, J. Van Orshoven, B. Muys, H. Kros, G. W. Heil, and W. Van Deursen, "AFFOREST sDSS: a metamodel based spatial decision support system for afforestation of agricultural land," New Forests, vol. 30, no. 1, pp. 33–53, 2005.
21. L. Verchot, R. Zomer, O. Van Straaten, and B. Muys, "Implications of country-level decisions on the specification of crown cover in the definition of forests for land area eligible for afforestation and reforestation activities in the CDM," Climatic Change, vol. 81, no. 3-4, pp. 415–430, 2007.
22. B. Muys, J. Hynynen, M. Palahi, et al., "Simulation tools for decision support to adaptive forest management in Europe," Forest Systems, vol. 19, no. 1, pp. 86–99, 2011.
23. T. Eid and K. Hobbelstad, "AVVIRK-2000: a large-scale forestry scenario, model for long-term investment, income and harvest analyses," Scandinavian Journal of Forest Research, vol. 15, no. 4, pp. 472–482, 2000.
24. The Stern Review, "The Economic of Climate Change," 2006,http://www.icaew.com/en/library/subject-gateways/environment-and-sustainability/stern-review/.
25. M. Nijnik, "Carbon capture and storage in forests," in Carbon Capture: Sequestration and Storage, R. E. Hester and R. M. Harrison, Eds., vol. 29 of Issues in Environmental Science and Technology, pp. 203–238, The Royal Society of Chemistry, Cambridge, UK, 2010.
26. F. Schmithüsen, "Multifunctional forestry practices as a land use strategy to meet increasing private and public demands in modern societies," Journal of Forest Science, vol. 53, no. 6, pp. 290–298, 2007.
27. G. C. van Kooten, Climate Change Economics, Edward Elgar, Cheltenham, UK, 2004.
28. M. Nijnik and L. Bizikova, "Responding to the kyoto protocol through forestry: a comparison of opportunities for several countries in Europe," Forest Policy and Economics, vol. 10, no. 4, pp. 257–269, 2008.
29. D. Anderson, "Economic aspects of afforestation and soil conservation projects," The Annals of Regional Science, vol. 21, no. 3, pp. 100–110, 1987.

30. H. Phillips, Secure Resource Supply for the European Wood Industry, Inventory scenarios for timber resources in Ireland. COST E44 workshop, Dublin, Ireland, 2006.
31. M. Michetti and R. Nunes, Afforestation and Timber Management Compliance Strategies in Climate Policy. A Computable General Equilibrium Analysis, vol. 2011.004 of FEEM Note di lavoro, 2011.
32. M. Lindner, M. Maroschek, S. Netherer et al., "Climate change impacts, adaptive capacity, and vulnerability of European forest ecosystems," Forest Ecology and Management, vol. 259, no. 4, pp. 698–709, 2010.
33. G. Nabuurs, R. Paivinen, and H. Schanz, "Sustainable management regimes for Europe's forests—a projection with EFISCEN until 2050," Forest Policy and Economics, vol. 3, no. 3-4, pp. 155–173, 2001.
34. G. Nabuurs, R. Paivinen, M. J. Schelhaas et al., "Nature-oriented forest management in Europe: modeling the long-term effects," Journal of Forestry, vol. 99, no. 7, pp. 28–34, 2002.
35. A. Shvidenko and Andrusishin, Ukraine: The Conditions and Prospects of the Forest Sector, IIASA, Laxenburg, Austria, 1998, Unpublished manuscript.
36. State Committee of Forestry of Ukraine (SCF), State Records. Kyiv, 2007, http://www.gov.org.ua.
37. I. Soloviy, "New forests for Europe: afforestation at the turn of century," in Proceedings of the EFI, vol. 35 of Afforestation in Ukraine—potential and restrictions, pp. 195–211, February 2000.
38. A. Strochinskii, Y. Pozyvailo, and S. E. Jungst, "Forests and forestry in Ukraine: standing on the brink of a market economy," Journal of Forestry, vol. 99, no. 8, pp. 34–38, 2001.
39. S. Gensiruk and M. Nijnik, Geography of Forests in Ukraine, Lviv Publisher, L'viv, Ukraine, 1995.
40. M. Nijnik and G. C. Van Kooten, "Forestry in the Ukraine: the road ahead?" Forest Policy and Economics, vol. 1, no. 2, pp. 139–151, 2000.
41. The President's Decree aimed to reform the forestry sector of Ukraine, 2004, http://www.gov.org.ua.

42. The State Programme, "Forests in Ukraine, 2002–2015," Kyiv, Ukraine, 2002, http://www.rada.gov.ua.
43. The Forest Code of Ukraine, 2006, http://zakon.rada.gov.ua/cgi-bin/laws/anot.cgi?nreg=3852-12.
44. M. Nijnik, To sustainability in forestry: the Ukraine's case, Ph.D. thesis, Wageningen University, 2002.
45. FAO, Forstat, 2010, http://www.fao.org/forestry/country/en/ukr/.
46. Ministry of Environment and Natural Resources (MENR), "The National Report. Kyiv," 2003,http://myland.org.ua/index.php?id=&lang=en.
47. National Academy of Sciences (NAS), Ukraine: the Prognosis for the Development of Production Forces, Kyiv, 2002.
48. R. Bun, M. Gusti, L. Kujii , et al., "Spatial GHG inventory: analysis of uncertainty sources. A case study for Ukraine," in Accounting for Climate Change: Uncertainty in Greenhouse Gas Inventories—Verification Compliance and Trading, D. Lieberman, M. Jonas, Z. Nahorski, and S. Nilsson, Eds., pp. 63–74, Springer, Amsterdam, The Netherlands, 2007.
49. S. Nilsson and A. Shvidenko, "The Ukrainian Forest Sector in a Global Perspective," IR-99-011/March. IIASA, Laxenburg, Austria, 1999.
50. N. Hanley and C. Spash, Cost-Benefit Analysis and the Environment, Edward Elgar, 1993.
51. M. Nijnik, "Economics of climate change mitigation forest policy scenarios for Ukraine," Climate Policy, vol. 4, no. 3, pp. 319–336, 2004.
52. G. Van Kooten and E. Bulte, The Economics of Nature: Managing Biological Assets, Blackwell Publications, Oxford, UK, 2000.
53. M. Nijnik, "To an economist's perception on sustainability in forestry-in-transition," Forest Policy and Economics, vol. 6, no. 3-4, pp. 403–413, 2004.
54. P. Lakida, S. Nilsson, and A. Shvidenko, "Estimation of Forest Phytomass for Selected Countries of the Former European USSR," WP-95-79, IIASA, Laxenburg, Austria, 1995.
55. A. Jessome, "Strength and related properties of woods grown in Canada. Forestry," Tech. Rep. 21, Ottawa, Canada, 1977.

56. R. Sandor and J. Skees, "Creating a market for carbon emissions opportunities for US farmers," Choices Quarter, vol. 1, pp. 13–18, 1999.
57. S. Huq and H. Reid, "Mainstreaming adaptation in development. IDS bulletin," Climate Change and Development, vol. 35, pp. 15–21, 2004.
58. D. Klooster and O. Masera, "Community forest management in Mexico: carbon mitigation and biodiversity conservation through rural development," Global Environmental Change, vol. 10, no. 4, pp. 259–272, 2000.
59. K. Naka, A. L. Hammett, and W. B. Stuart, "Forest certification: stakeholders, constraints and effects,"Local Environment, vol. 5, no. 4, pp. 475–481, 2000.
60. E. Fernandez, L. Rojo, M. N. Jimenez, et al., "Afforestation improves soil fetiliyu in south-eastern Spain," European Journal of Forest Research, vol. 194, no. 4, pp. 707–717, 2010.
61. M. Nijnik, B. Slee, and G. Pajot, "Opportunities and challenges for terrestrial carbon offsetting and marketing, with some implications for forestry in the UK," South-East European Forestry Journal, vol. 1, no. 2, pp. 69–79, 2011.
62. S. Fankhauser and L. Lavric, "The investment climate for climate investment: joint Implementation in transition countries," Climate Policy, vol. 3, no. 4, pp. 417–434, 2003.
63. E. Petkova and G. Faraday, "Good Practices in Policies and Measures for Climate Change Mitigation," REC and WRI, Szentendre, Hungary, 2001.
64. IBN-DLO, Resolving Issues on Terrestrial Biospheric Sinks in the Kyoto Protocol. Dutch National Research Programme on Global Air Pollution and Climate Change, Wageningen, The Netherlands,1999.
65. R. N. Lubowski, A. J. Plantinga, and R. N. Stavins, "Land-use change and carbon sinks: econometric estimation of the carbon sequestration supply function," Journal of Environmental Economics and Management, vol. 51, no. 2, pp. 135–152, 2006.
66. G. Van Kooten, Land Resource Economics and Sustainable Development. Economic Policies and the Common Good, UBC Press, Toronto, Canada, 1993.

67. J. Lampietti and J. Dixon, To See the Forest for the Trees: A Guide to non-Timber Forest Benefits, World Bank, Washington, DC, USA, 1995.
68. P. Lakyda, I. Buksha, and V. Pasternak, "Opportunities for fulfilling Joint Implementation projects in forestry in Ukraine," in Forest, Climate and Kyoto, vol. 222 of Unasylva, pp. 1–6, 2005.

Chapter 6

Forest Observational Studies-An Essential Infrastructure for Sustainable Use of Natural Resources

XiuHai Zhao[1], Javier Corral-Rivas[2], ChunYu Zhang[1], Hailemariam Temesgen[3], and Klaus v Gadow[4,5]

[1]Beijing Forestry University, Beijing, China

[2]Universidad Autonóma de Nuevo León, Durango, México

[3]Oregon State University, Corvallis, USA

[4]Burckhardt Institute, Georg-August University, Göttingen, Germany

[5]Institute of Forestry and Wood Science, University of Stellenbosch, Stellenbosch, South Africa

ABSTRACT

This contribution complements *Forest Ecosystems'* Thematic Series on "Forest Observational Studies". We provide essential clarification regarding the definition and purpose of long-term field studies, review some of the extensive literature and discuss different approaches to collecting field data. We also describe two newly established forest observational networks that serve to illustrate the scope and diversity of forest field studies. The first is a large-scale network of forest observational studies in prominent natural forest ecosystems in China. The second example demonstrates observational studies in mixed and uneven-aged pine-oak forests which are selectively managed by local communities in Mexico. We summarize the potential for analysing and modeling forest ecosystems within interdisciplinary projects and provide argumentation in favour of long-term institutional commitment to maintaining forest observational field studies.

REVIEW

This section clarifies definitions and reviews some of the extensive literature relating to forest resource assessments, manipulated experiments and observational studies in forest ecosystems. Some parts were previously presented in two workshop proceedings that had not been subject to peer review (Zhao et al. 20112012; Amateis and Burkhar 2012).

Forest Resource Assessments (Forest Inventories)

The objective of a resource assessment, such as a National Forest Inventory (NFI), is to provide unbiased estimates of particular target variables and to present statistics about forest resources. NFIs assess forest areas, growing stock volumes, and changes in biodiversity status, land-use, carbon stock and ecosystem services. International agreements, including the Montréal Process (1998), the Ministerial Conference on the Protection of Forests in Europe (2008) and the Convention on Biological Diversity (2008), require information which is provided by NFIs. In North America, the Forest Inventory and Analysis (FIA) Program of the U.S. Forest Service provides the information needed to assess America's forests (LaBau et al. 2007). Naturally, sampling designs, sample plot configurations and other assessment methods may vary among countries and there is diversity in the definitions of forest area and growing stock volume among countries. Thus, harmonizing assessment and reporting is essential to make NFI results comparable (Tomppo et al.2010).

Manipulated Experiments

A *manipulated experiment* is an investigation that attempts to establish a particular set of conditions under a specified protocol with the aim of testing a hypothesis. The adjective*manipulated* implies the establishment of a set of predefined treatments which allow comparison of the effects/responses resulting from these treatments (Fisher 1935; Cox 1958; Gadow and Kleinn2005). Thus, the experiment deliberately imposes a treatment on a group of objects in the interest of observing the response. Examples of manipulated experiments include medium and long-term growth studies in response to different fertilizer applications and stand densities (O'Hehir 2001; Burkhart and Tomé2012), growth studies of clones on different sites, including marginal ones (Bungart and Hüttl 2004), growth and competition effects in multi-species forests (Vanclay 1994; Pretzsch 2009), or evaluating effects of afforestation on water yield in mountain catchments (Bosch and Gadow 1990). Despite attempts to create homogenous conditions, manipulated field experiments always involve uncertainties in controlling *ceteris paribus* conditions, which are necessary for obtaining noise-free "dose/ response" relations.

Manipulated field experiments do not seem to be as popular as they used to be during the 1960's and 70's. Possible reasons are the high cost of maintaining the field experiments and the restricted ability to generalize beyond the homogeneous, and therefore limited, experimental conditions. An exception is the impressive network of the Forest Modeling Research Cooperative at Virginia Tech which has employed a combination of permanent sample plots in operational stands of loblolly pine (*Pinus taeda*) plantations and designed experiments to provide data bases needed to construct robust growth and yield models for projecting inventories and estimating response to a wide range of silvicultural treatments for plantations established primarily for wood production (Burkhart2008). These extensive networks have provided the empirical base for models of tree taper and volume, site index, and tree and stand increment and mortality. Data from spacing trials that were measured annually over a 25-year period (Amateis and Burkhart 2012) and from silvicultural trials designed for estimating response to control of competing vegetation and fertilizer applications extend and enhance the basic modeling framework. (Amateis and Burkhar 2012). Because the plots were stem-mapped and measured regularly, it was possible to obtain data on decay rates and amounts of carbon in the dead trees at the time plot measurements were scheduled to be terminated in the first region-wide study (Radtke et al. 2009). With increased interest in climate influences on forests, both region-wide data sets, which cover a broad range of soil and climate zones where loblolly pine is planted, are providing highly valuable data for a large multi-disciplinary, multi-institutional research effort aimed at assessing climate influences on southern U.S. conifers (http://www.pinemap.org).

Forest Observational Studies

Models of forest ecosystem dynamics, including tree growth, mortality and recruitment, are often developed on the basis of repeated observations in specially selected observational field plots. Selection of sites is not random or systematic, as in forest inventories, but based on particular data requirements. Ecologists tend to use the term "opportunistic sampling" for this approach (De Barba et al. 2010). The aim of the early field experiments established during the 19th century was to measure the growth of trees which were numbered for re-identification during successive measurements. Some of these experiments have been maintained for over a century, providing valuable information on long-term developments (Spellmann et al. 1996). The field plots are usually larger than inventory plots, and measurements are often much more detailed. Despite advances in resource assessment technology, there seems to be a real need for long-term observational field studies with mapped trees in large field plots. Examples of the design and implementation of such studies are presented by Zhao et al. (2011), Corral-Rivas et al. (2012), Tewari et al. (2014) and Kiviste et al. (2012).

In contrast to a *manipulated* experiment, which deliberately imposes treatments on experimental plots with the aim of observing a particular response, a *comparative observational study* involves collecting and analysing data from different site conditions without actively pre-defining these conditions (Kuehl 1994). Comparative observational studies are also known as *quasi-experiments*(Campbell and Stanley 1963; Cook and Campbell 1979). Typical *quasi-experiments* are longterm forest observational studies. Extensive networks of longterm forest observational studies have been established since the middle of the nineteenth century. Franz v. Baur (1830–1897), Bernhard Danckelmann (1831–1901) and Gustav Heyer (1826–1883) were among the first who devised a concept for long-term investigations in forest science, emphasising the importance of experimental field plots. In 1929 the *International Union of Forest Research Organisations* (IUFRO) was established with the objective of standardising the design and analysis of long-term field experiments (Pretzsch 2009). The majority of these experiments are observational studies (Szaro et al. 2006). Numerous observational studies were also established in North America. (Seymour et al. 2006a) reviewed long-term silvicultural experiments in four regions of the United States. As early as the 1920s, when there was much interest in multi-aged silviculture, scientists recognized that silvicultural systems involving within-stand variation of tree age and size could not be tested effectively on small (<1 ha) plots, and began installing compartment-scale (10–20 ha) trials on many experimental forests throughout the United States. Such large-scale trials have experienced a revival during the 1990's in response to a renewed interest in management methods aiming to maintain within-stand structural complexity and biodiversity at larger scales.

Longitudinal, Cross-Sectional and Interval Studies

Forest ecosystem studies are often designed as *longterm* experiments, which are known in the statistical literature as *longitudinal studies*. The key feature of longitudinal data is the fact that the same individual is repeatedly measured at successive points in time. Thus, the set of observations on one individual subject will tend to be positively correlated and this correlation needs to be taken into account (Crawley 2005, p. 180). Longitudinal studies are commonly used to describe and explain trends. One disadvantage of a longterm or "permanent" experiment is the high maintenance cost of the research infrastructure and the long wait for results (Zhao et al. 2011).

On the other hand, results are available relatively quickly in a *cross-sectional study* (also known as a *chronosequence*) which involves one-time measurements of a set of field plots which usually cover a wide range of ages and environmental conditions. Thus, the sequence of remeasurements in time (longitudinal study) is substituted by simultaneous measurements in space. This method has been used extensively during the 19th century (Kramer 1988, p. 97). *Cross-sectional studies*may be combined with stem analyses to reconstruct the development of tree height, for example (Lee 1993). They may provide information relatively quickly, but do not capture the response of a target variable to a given initial state. The initial state may be defined for example, by the competition effects in the past. A *cross-sectional study* evaluates differences in growth in response to different site conditions. It does not provide evidence that can be used to test an effect such as past forest density or competition. A *cross-sectional study* is the only possible approach, however, in experiments requiring destructive sampling, i.e. in biomass studies where individual trees, or a cohort of trees within a given area, have to be cut up, dried and weighed. Obviously, the selection of the particular individuals or the particular cohorts will greatly influence the results. A *cross-sectional study* provides one-time measurements for many individuals, but no rates of change. A*longitudinal study* does assess rates of change, but usually only for few individuals. Thus, there are two extreme cases (Crawley 2005, p. 180):

1. 1. few measurements on many individuals
2. 2. many measurements on few individuals

Cross-sectional studies typically involve few observations on a large number of individuals while in*longitudinal studies* few individuals are typically measured many times. A practical compromise between a *longitudinal study* and a *cross-sectional study* is an *interval study*. *Interval* plots are measured at least twice with the objective of assessing the rate of change of a target variable between successive re-measurements. As many sites as possible should be covered and there is no immediate need to observe the target variable for long periods of time. Thus *interval studies*strike a compromise, exploiting the primary advantages of *longitudinal experiments* (gathering specific age and cohort-related responses) and *cross-sectional studies* (broad coverage of site conditions and minimum wait for data). However, continuous observations over long time periods are indispensable for assessing certain longterm effects, such as maximum density and tree survival for different ages and initial espacements, and the effects of changing environmental conditions.

An important question concerns the balance between the number of growth intervals that should be assessed in one particular location and the range of environmental and treatment conditions that needs to be covered. In the ideal case, one growth interval will provide sufficient response and the funds that would be required for assessing a second interval at the particular location can be spent on gathering another set of interval data in a different locality. An interval plot should be continued after one growth period if new results are expected, or it may be abandoned and the available funds used in another location to increase the variety of initial states for which the response needs to be evaluated. Generally, interval plots are continued for multiple growth periods, but ongoing evaluation should be carried out of the tradeoff between abandoning current interval plot locations and establishing new plots in additional locations to increase the variety of initial states for which response data are needed.

Some plots have been continuously re-measured over long periods of time. Pelz and Kohnle (2012) list numerous examples of long-term field studies in Germany. New networks of such studies, some rather extensive, were established more recently. The impressive examples from the US, China and Mexico are described below in more detail. However, there are also examples of observational studies that had been established during the early years of the 20th century, were subsequently abandoned (usually after political change caused by a revolution or independence) and are being re-initiated. One such series was described by Sims et al. (2009). Concerted efforts are also being made in India to locate and revive old plots which had been established during colonial times (Tewari et al. 2014).

The ultimate aim of all field studies is to provide external validity, the ability to generalize from a limited set of observations. Generalizability depends on whether the observed response measurement is representative of a population of interest. Scientists need to clarify whether the results of the observations may be legitimately extended to the general population of interest. In theory, it should be possible to extend observations beyond the particular restricted data set. However, this is not always possible in practice, for example in the case of rare biotopes or endangered species that occur in geographically restricted areas. A comprehensive description of the study sites and of the methodology is helpful for users to judge whether the results are applicable to a particular situation.

Resource Assessments, Manipulated Experiments and Observational Studies

Thus, according to Zhao et al. (2011) we may distinguish three common approaches for collecting field data in forest ecosystems: Resource Assessments, Field Experiments and Observational Studies. *Resource Assessments* are usually carried out at regular intervals, providing geographical information about specific target variables (biomass; tree species). The data are used for formulating policy by governments, land managing agencies and NGOs, for forest planning, and for locating timber resources and new processing facilities. A *manipulated experiment* deliberately imposes treatments on experimental plots with the aim of observing a particular response. Field experiments are more often designed for planted forests with the aim of developing optimum silvicultural treatments. *Observational studies* are often laid out in natural forests with irregular structures where treatments are difficult to define. Data are collected from different site conditions without actively pre-defining these conditions. Field plots are usually large to capture effects of scale, and assessments are scheduled to coincide with a disturbance. Resource Assessments, Field Experiments and Observational Studies may complement each other. However, in *Observational Studies* the primary objective is not hypothesis testing or geographical coverage, but improved understanding of ecosystem dynamics, especially regarding the immediate and long-term effects of harvest events and other disturbances.

Numerous long-term observational studies with small, large and nested plot structures have been installed and maintained by various institutions in different countries covering different time periods and diverse forest ecosystems. The utility of each design depends on specific data requirements (forest trees, or other vegetative life forms) and available funds. The following two examples represent successful new implementations.

EXAMPLES

This section briefly presents details of the development and current status of two large observational networks, the Beijing observational network in China and the Durango observational network in Mexico. The two studies provide information about forest dynamics in multi-species, uneven-aged forests which are unmanaged (China) and multi-species, uneven-aged forests which are subject to selective harvesting (Mexico). As such, these studies are representative of unmanaged and managed (near-) natural ecosystems.

The Beijing Forest Observational Network

The first three field plots of the Beijing Forest Ecosystems Observation Network were established in 2005 in the Changbaishan forest region (plots 3a, 3b and 3e in Table 1). Two of these plots had been selectively cut by the Forestry Bureau in 1998. The third plot had been left unmanaged. Establishing managed and unmanaged field plots in close vicinity to each other has since become one of the guiding principles of the network. Standard methods of assessment were also defined at that time. During the past decade, the network has expanded to include all of the most important natural forest ecosystems in China (Figure 1).

Table 1: Some details of the field research clusters

No.	English name	Chinese name	Type	Near	Area (ha)	Forest type	1st meas	2nd meas
1	Liangshui	凉水	U	NA	30.0	*Md*	2010	-
2a	Jiaohe UN	蛟河UN	U	V	10.0	*Md*	2010	-
2b	Jiaohe UF	蛟河UF	U	NE	20.0	*Md*	2010	-
2c	Jiaohe UN	蛟河UN	U	F	21.1	*Md*	2010	-
2d	Jiaohe CRN	蛟河CRN	CR	V	10.0	*Md*	2010	-
2e	Jiaohe CRF	蛟河CRF	CR	NE	20.0	*Md*	2010	-
2f	Jiaohe UN	蛟河UN	CR	F	21.1	*Md*	2009	-
2 g	Jiaohe UA	蛟河UA	U	NA	30.0	*Md*	2010	-
3a	Changbaishan I CMNa	长白山I CMNa	CM	V	5.2	*Md*	2005	2010
3b	Changbaishan I CMNb	长白山I CMNb	CM	F	5.2	*Md*	2005	2010
3c	Changbaishan I UF	长白山I UF	U	NA	5.2	*Md*	2007	-
3d	Changbaishan I UNa	长白山I UNa	U	NE	1.4	*Md*	2006	-
3e	Changbaishan I UNb	长白山I UNb	U	NA	1.0	*Md*	2005	2010
4a	Changbaishan II UV	长白山II UV	U	NA	5.2	*Pin/Ab*	2008	-
6	Keqi	克旗	U	NA	4.0	*Pinus*	2008	-
7	Songshan	松山	U	NA	1.2	*Pinus*	2007	-
8	Heilihe	黑里河	U	NA	4.0	*Pinus*	2007	-
9	Taiyueshan	太岳山	U	??	4.0	*Pinus*	2007	-

10	Helanshan	贺兰山	U	NA	10.0	*Pinus*	planned	-
11	Wanglang	王朗	U	NA	10.0	Pic/Ab	planned	-
12	Jiuzhaigou	九寨沟	U	NA	4.0	Pic/Ab	2007	-
13	Jiuzhaigou	九寨沟	U	NA	4.0	*Pinus*	2007	-
14	Laojunshan	老君山	U	NA	4.0	*Pic/Ab*	2008	-
15	Yulongxueshan	玉龙雪山	U	NA	4.0	*Pic/Ab*	2008	-
16	Linzhi	林芝	U	NA	4.0	*Pic/Ab*	planned	-
17	Qilianshan	祁连山	U	NA	10.0	*Pinus*	2010	-
18	Tianshan	天山	U	NA	4.0	*Picea*	2007	-
19	Jiulianshan	九连山	U	NA	5.0	*BE*	2009	-

V, F, NE and NA refer to the assumed "nearness" to a natural state (V = very far; F = far; NE = near; NA = nature); U, CM and CR refer to the treatment (U = unmanaged; CM = selective harvesting by management; CR = selective harvesting by research). A is an "attractor" plot, which represents a guiding model for management. Md = mixed deciduous; P/A = Picea/Abies; Pin/Ab = Pinus/Abies; BE = Broadleaved Evergreen.

Zhao *et al.*

Zhao *et al. Forest Ecosystems* 2014 **1**:8, doi:10.1186/2197-5620-1-8

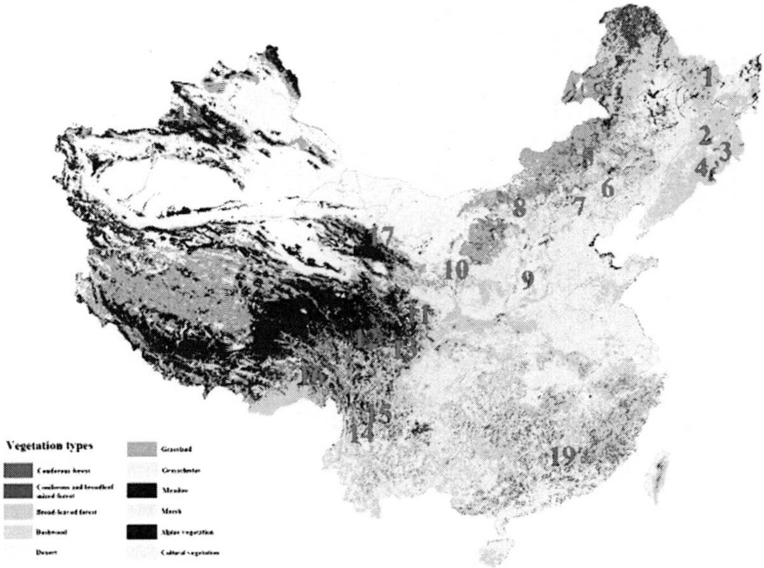

Figure 1: Geographical distribution of field plots and plot clusters in the Beijing forest ecosystems network.

Some details of the field plots and plot clusters are presented in Table 1. The network includes managed and unmanaged experiments. Both are designed to study ecosystem structure and dynamics, including tree growth, recruitment and mortality. The focus in the unmanaged series is on a wide range of topics that are of interest to ecologists, such as species-habitat associations; gender-related biomass partitioning; natural ecosystem structure and diversity; response to natural disturbances such as wildfire; seed dispersion and seedling survival.

The development in the managed experiments is characterized by regular human disturbance through selective harvesting. Thus, the major objectives in the managed series is to assess and analyse modifications of ecosystem structure and diversity caused by harvest events, to study forest dynamics under management and to develop practical guidelines for selective harvesting under continuous cover forestry. Assessments are not limited to measuring trees. Flowers and fruits of dioecious tree and shrub species are collected, tree ring data assessed, seeds and seedling survivals counted, biomass components sampled and the response of the leaf spectral reflectance to above- and belowground competition is studied under field conditions.

Intensive research started immediately after the plot installations and first results were published after suitable datasets became available (Zhang et al. 2009a; Zhang et al. 2009b; Gao et al. 2010; Zhang et al. 2010a; Zhang et al. 2010b; Liu et al. 2011; Wang et al. 2011). The future objectives of the network include the need to secure long term institutional support, the desire to attract scholars from a variety of disciplines, a focus on harvest event analysis and continuing research on ecosystem structure and change.

Maintenance and extension of the new forest observational network is currently secured for periods up to 5 years in accordance with the five-year plans of the People's Republic of China (Chinese: 中国五年计划; Pinyin: Zhōngguó Wǔnián Jìhuà) and its economic development guidelines. Additional support is secured from time to time by short-term projects extending over periods of 1–2 years. Because of the enormous size of the network and the country-wide coverage of virtually all the important forest ecosystems, the network requires institutional support to ensure the survival of this unique "Green Infrastructure", and the establishment of several observational headquarters in different parts of the country is currently being discussed.

The Durango Forest Observational Network

The forest ecosystems of Durango, covering approximately 5 million hectares, represent the most important timber resource in Mexico. These forests are rich in biodiversity with at least 27 coniferous tree species, among them 20 species of *Pinus,* and 43 species of *Quercus*. The predominant forest types are mixed and uneven-aged pine-oak forests, representing more than two-thirds of the total area. The forests of Durango have been managed by local communities for more than 70 years, mainly using selective harvesting, for sustainable timber production but also for the maintenance of biological diversity and uneven-aged stand structures. Such selective forest management requires regular monitoring, and a new network of observational studies has been established. The network comprises 429 field plots, covering 0.25-ha each, that were established between 2007 and 2011, on the Sierra Madre Occidental (Figure 2).

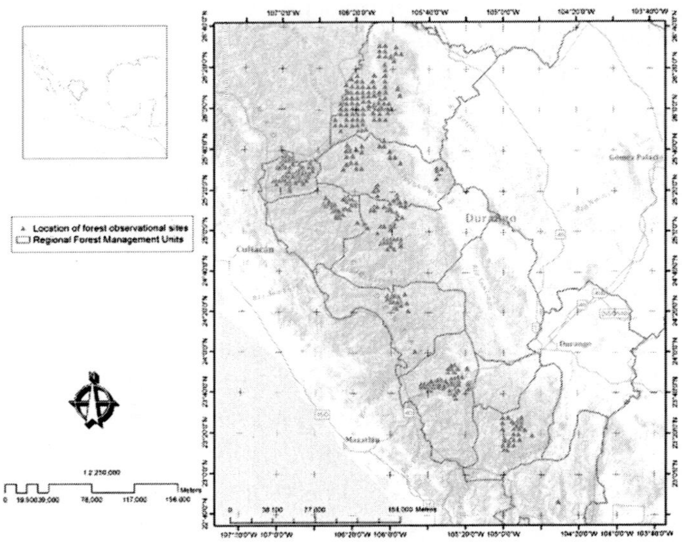

Figure 2: Location of the permanent forest observational sites in the pine-oak forests of Durango and some statistics of the 429 (50 × 50m) fiel plots of the Durango observational of the 429 (50 × 50m) fiel plots of the Durango observational network. The number of tree species recorded within the plots vary from 1 to 13 (average 7).

Table 2 summarizes the stand variables calculated from the tree data assessed in the observational studies. The maximum and minimum values of the stand variables indicate that the network adequately represents the stand variability of mixed and uneven-aged forests of Durango. For example, the density of the sites ranges from 3.11 to 43.30 basal area per hectare. Such a wide amplitude of densities will make it possible to identify ideal residual (retention) basal areas to promote better rates of biomass accumulation.

Table 2: Summary of selected stand variables of the Durango Observational Network: trees per hectare (N), stand basal area (G, m^2 ha^{-1}), quadratic mean diameter (dg, cm), mean breast height diameter (d, cm), mean total height (h, m), dominant height (H_0, m), and dominant diameter (D_0, cm)

Stand variable	*N (N/ha)*	*G (m^2/ha)*	*Dg (cm)*	*d (cm)*	*h (m)*	*H_0 (m)*	*D_0 (cm)*
Mean	661.98	19.99	20.46	18.00	11.09	18.97	56.04
Maximum	1600.00	43.28	45.46	42.31	23.21	36.60	95.10
Minimum	64.00	3.13	12.35	11.23	5.03	5.40	15.86
Standard deviation	324.46	7.96	4.84	4.49	3.35	6.98	16.38

The last two variables were calculated as the top height and mean of the 100 largest diameter trees per hectare.

Zhao *et al.*

Zhao *et al.* *Forest Ecosystems* 2014 **1**:8, doi:10.1186/2197-5620-1-8

The results of forest spatial analyses show that most tree species exhibit a high degree of spatial mingling;

that the spatial distribution of small and large trees is clustered; and that horizontal and vertical forest structures are very irregular. Observed spatial tree mark associations seem to be highly linked with seed dissemination processes, the relative shade tolerance of some tree species, and specific preferences for particular soil types. On the other hand, repulsive interactions could be an indication of competition processes that impose minimum distances between marks of trees. Selective tree removal is a major cause of spatial structuring (Longuetaud et al. 2008). For this reason, balanced structure areas are defined as the minimum contiguous area that is required for sustainable management of a multi-sized selection forest (Wehenkel et al. 2011). Such minimum areas which exhibit a balanced diameter distribution, and the corresponding number of trees, for different tree species and forest types are estimated based on data obtained from the new observational network.

Maintenance and extension of this forest observational network is currently secured for the next 6 years. The first 93 sites were re-measured during the winter of 2012. Mexico's National Forestry Commission recently approved the funding of 870 new observational study sites in Durango and an additional 3000 sites in other Mexican States. This network has thus already developed into a "green" infrastructure of national importance.

DISCUSSION AND CONCLUSIONS

In this section, we show how forest observational studies may become part of a country's infrastructure. We also summarize the potential of long term field studies for analysing and modeling forest ecosystems in the context of interdisciplinary research. Finally, we discuss potential synergies between Forest Observational Studies and other Data Sources.

Forest Observational Studies - Ideal Platform for Interdisciplinary Research

An important longterm objective of many forest observational studies is to attract scholars from a variety of disciplines to provide holistic and hierarchically structured information about natural systems (including planted forests), and to exploit the potential for interdisciplinary research. Interdisciplinary research attempts to integrate methods, concepts and data from two or more disciplines (Ausburg 2006). Such efforts may advance fundamental understanding or help to solve problems whose solutions are beyond the scope of a single discipline. According to Kafatos and Eisner (2004), many of the great research triumphs are products of interdisciplinary inquiry and collaboration, including the discovery of the structure of DNA, magnetic resonance imaging, and human genome sequencing.

There are numerous examples of large scale ecosystem field studies which have been designed for interdisciplinary inquiry. Monserud (2002) describes seven silvicultural studies in the moist-maritime coastal regions of the Pacific Northwest: the *Alternatives to Clearcutting* study in Alaska; the *Montane Alternative Silvicultural Systems* in British Columbia; the *Olympic Habitat Development Study*, the *Forest Ecosystem Study* and the *Washington DNR Capitol Forest Study* in Washington; the *Demonstration of Ecosystem Management* study in Washington and Oregon and the *Density Management Study* in Oregon. These seven interdisciplinary field studies are characterized by large spatial units with replicated treatments and multiple objectives (Szaro et al.2006). The maintenance of such large experiments tends to be rather expensive, however, which may shorten their lifespan due to lack of funding, and thus restrict the essential continuity required in forest observational studies. Nevertheless, these studies have shown that field experiments provide an ideal platform for interdisciplinary investigation.

Using Synergies between Forest Observational Studies and Other Data Sources

Data from observational experiments collected over long time periods may effectively complement forest inventories. New longterm observational networks with large plot areas and mapped tree data in America, Asia and Europe offer improved opportunities for detailed analysis, especially in uneven-aged, species-rich forests. Growth models are usually developed using data from long term FOS, but the spatial representation is often restricted. In order to test the applicability of the particular growth model for other regions, these models may be calibrated with data from forest inventories.

Pelz and Kohnle (2012) cite (Lessard and Van Deusen Paul 2001) who states that growth models may be used to update information on inventory plots not visited in a particular year and that in turn, diameter growth models may be calibrated with inventory data by means of residual analysis. A similar calibration approach was reported by Nothdurft and Schmidt (2010). (de Menezes et al. 2000) list several methods for combining FOS and inventory data sets.

Questions of Scale

Regarding spatial scale, (Seymour et al. 2006b) found that large-scale studies are`difficult and expensive to install, replicate, monitor, and maintain over time. But there are also important benefits. Scientists learn to appreciate operational realities of forest management, and study sites provide realistic demonstrations of new silvicultural systems. Plot installations offer field laboratories where a wide range of questions can be addressed, including small-scale phenomena, such as amphibian dispersal and seedling development and whole-stand responses to silvicultural treatment. Large study areas are essential for developing species-area relationships (Hubbell2001) where the effect of increasing areas on the number of species can be studied on the same site (Gadow and Hui 2007). Studies of competition effects at different spatial scales, species' habitat preferences (Zhang et al. 2009b) and analyses of sexual dimorphism in a dioecious tree species with widely scattered individuals (Zhang et al. 2012) also require large field plots.

Long-term studies are required for assessing the maximum density of a forest ecosystem. A classic example is provided by the South African "Correlated Curve Trend" (CCT) studies. The CCT studies were established with a wide range of extremely low and high densities. Some plots have been observed for more than 70 years providing unique information about tree survival in response to different initial spacings (Dickel et al. 2010). Trees are long-living organisms which react differently to changing environmental conditions during different life stages and long-term studies are required to assess these reactions.

It is not possible to capture all ecosystem phenomena and to apply all treatments of interest to a single network of permanent plots. Consequently, data from a variety of installations are often used in combination to develop comprehensive modeling frameworks (Amateis and Burkhart2012). Thus forest modeling often helps to identify knowledge gaps, and to integrate and synthesize information from a variety of sources, to provide sound forecasting capability.

New Technical and Methodological Advances

Three dimensional terrestrial laser scanning (TLS) and airborne laser scanning (ALS) possibly represent the most important technical advances in measurement and monitoring techniques relevant to forest observational studies. Ground-based TLS is used to measure objects in three dimensional space by emitting a laser beam and scanning the surface of surrounding objects using impulse or phase-based systems (Seidel 2014). Terrestrial laser scanning provides data about tree position, height, diameter at breast height, crown dimension, as well as tree lean, sweep and taper. However, suitable software for analysing the huge amounts of raw data does not seem to be freely available yet. ALS is based on similar technical principles, but the scanner is usually mounted on a small aircraft or helicopter. Possibly the first comprehensive textbook about forest applications of airborne laser scanning is the work by Maltamo et al. (2014) which provides a comprehensive, state-of-the-art review of the research and application of ALS in a broad range of forest-related disciplines. Some chapters deal with methodological detail while others review and illustrate the use of ALS within various domains of application to forest ecosystems.

New methods of analysing forest ecosystem structure and dynamics are being developed continuously. Examples include the ability to partition temperate plant community structure at different scales, to analyse spatial distributions and spatial associations of dominant tree species (Zhang et al. 2009a; Zhang et al. 2010b) and to describe gender-related distributions, competition and habitat effects in old-growth forests (Zhang et al. 2009b; Zhang et al. 2010a). These methods are particularly useful and revealing if plot sizes are large to allow analysing the effects of scale.

More recent examples of analysing and modeling forest ecosystem dynamics are also based on observational field studies. Using large observational study in Northeast China, Temesgen et al. (2014) used a nonlinear mixed effect model with species indicators to make tree height predictions compatible and precise for 23 tree species found in multi-layered forests in NE China. Corral-Rivas et al. developed generalized Height-Diameter Models with Random Parameters for Mixed, uneven- aged Forests in North-Western Durango, Mexico. Such models are important for estimating forest biomass. Laiho et al. (2014) estimate the height increment of understorey trees under different canopies while Valkonen et al. (2014) estimate ingrowth and survival of those trees. Both studies provide essential input for evaluating forest treatment options (Pukkala 2014; Vanclay 2014). Such modeling is only possible with appropriate measurements, collected in observational studies, sometimes over extended time periods.

Timing Assessments to Analyse Disturbances

Models of tree growth, recruitment and survival represent the overwhelming result of traditional observational studies. But these models only describe a part of ecosystem dynamics. Equally important are the modifications caused by regular human disturbance in the form of clearfellings, plantings or tree harvesting. There is an abundance of tree growth models. In contrast, harvest event models predicting structural modifications based on silvicultural terminology, are virtually unknown. Yet such models are indispensable for comparing silvicultural alternatives in uneven-aged forests. To develop such a model requires harvest event data from observational studies where assessments are timed to coincide with the removals. A harvest event model should include estimates of the reduction of density (in terms of certain tree classes) and structural modifications (Mason et al. 2005). There is thus increasing motivation for analysing harvest events and using such data for modeling human disturbance.

Forest Observational Studies - An Essential Part of a Country's Infrastructure

The sustainable development of a country's forest ecosystems is the result of planned designs guided by a wide array of interests. Because they provide many essential services, forest ecosystems are generally considered a very important national asset. Data from observational experiments collected over long time periods may effectively complement forest inventory and monitoring. Growth models developed from observational studies are essential for projecting inventory data and for generating reasonable scenarios at landscape level. Long-term forest observational studies are therefore an essential part of a country's infrastructure, comparable to public transportation systems and energy supply services. Long term commitment and regular monitoring is required to ensure that forest resources are being sustained. Such commitment can only be guaranteed by the State.

An example of such a national infrastructure is given by Pelz and Kohnle (2012) who showed that at the end of the 19th century, forest research stations were established in Prussia, Baden, Saxony, Württemberg and Hesse with the objective of long term observation of selected forest types. In 1872, a formal cooperation among these research stations was established by the Association of forest research stations in Germany (a founding member of the International Union of Forest Research Organizations, IUFRO). Several of the observational plots that were initiated in the 19th century are still being re-measured today. The regular re-measurement, even during the war years, has produced an extensive and unique set of observations which provide the basis for detailed models of forest dynamics for different silvicultural treatments, and a variety of climatic and site effects. This achievement would not have been possible without the commitment of the State to create and continuously support such an extensive "green infrastructure". This infrastructure, staffed with permanently employed measurement teams, delivers essential information for monitoring national conservation efforts and develop climate adaptation strategies.

AUTHORS' CONTRIBUTIONS

All authors read and approved the final manuscript.

ACKNOWLEDGEMENTS

We greatly appreciate the constructive comments received from three anonymous reviewers which resulted in an extensive revision of the original submission. We thank the Forest Modeling Research Cooperative, and Harold Burkhart and Ralph Amateis at Virginia Tech, for providing material for this paper.

Individual author contributions: Zhao and Zhang (Section 2.2); Corral Rivas (Section 2.3); Temesgen (contributions to other sections): Gadow (Sections 1 and 3).

REFERENCES

1. Amateis RL, Burkhart HE: Rotation-age results from a loblolly pine spacing trial. *South J Appl For* 2012, 36:11-18.
2. Ausburg T: *Becoming Interdisciplinary: An Introduction to Interdisciplinary Studies*. 2nd edition. New York: Kendall/Hunt Publishing; 2006.
3. Bosch J, Gadow K: Regulating afforestation for water conservation in South Africa. *S Afr For J* 1990, No 153:41-54.
4. Bungart R, Hüttl RF: Growth dynamics and biomass accumulation of 8-year-old hybrid poplar clones in a short-rotation plantation on a clayey-sandy mining substrate with respect to plant nutrition and water budget. *Eur J For Res* 2004, 123(2):105-115.
5. Burkhart HE: Modelling growth and yield for intensively managed forests. *J For Sci* 2008, 24:119-126.
6. Burkhart HE, Tomé M: *Modeling Forest Trees and Stands*. Springer; 2012:457.
7. Campbell D, Stanley J: Experimental and Quasi-experimental Designs for Research and Teaching. In *Handbook on Research on Teaching*. Edited by Gage. Chicago: Rand McNally; 1963.
8. Convention on Biological Diversity: What is forest biological diversity? 2008. http://www.cbd.int/forest/what.shtml
9. Cook T, Campbell D: *Quasi-Experimental Design*. Chicago: Rand McNally; 1979.
10. Corral-Rivas JJ, Reyes RI, Wehenkel C, Aguirre-Calderón OA, Gadow K: *A Network of Forest Observational Studies in Durango (Mexico)*. Proceedings of an International Workshop at Beijing Forestry University; 2012:125-138.

11. Cox DR: *Planning of Experiments*. New York: Wiley; 1958.
12. Crawley MJ: *Statistics – An Introduction Using R*. Wiley; 2005:327.
13. De Barba M, Waits LP, Genovesi P, Randi E, Chirichella R, Cetto E: Comparing opportunistic and systematic sampling methods for non-invasive genetic monitoring of a small translocated brown bear population. *J Appl Ecol* 2010, 47:172-181. doi:10.1111/j.1365-2664.2009.01752.x
14. de Menezes LM, Bunn DW, Taylor JW: Review of guidelines for the use of combined forecasts. *Eur J Oper Res* 2000, 120:190-204.
15. Dickel M, Kotze H, Gadow K, Zucchini W: Growth and survival of *eucalyptus grandis* – a study based on modeling lifetime distributions. *Intl J Math Comput For Nat Resour Sci* 2010, 2(2):20-30. available online at HTTP://MCFNS.COM
16. Fisher RA: *The Design of Experiments*. 1st edition. London: Oliver & Boyd; 1935.
17. Gadow K, Kleinn C: Forest Management, Science-Based and Understandable. In*Balancing Ecosystem Values – Innovative Experiments for Sustainable Forestry*. Edited by Peterson CE, Maguire DA. US Dept of Agr. Forest Service, Gen. Tech. Rep. PNW-GTR-635; 2005:15-23.
18. Gadow K, Hui GY: Can the tree species-area relationship be derived from prior knowledge of the tree species richness? *For Stud* 2007, 46:13-22. 2007
19. Gao L, Zhang CY, Zhao XH, Gadow K: Gender-related climate response of radial growth in dioecious *fraxinus mandshurica* trees. *Tree-Ring Res* 2010, 66(2):2010.
20. Hubbell SP: *The Unified Neutral Theory of Biogeography and Biodiversity*. Princeton: Princeton University Press; 2001.
21. Kafatos FC, Eisner T: Unification in the century of biology. *Science* 2004, 303:1257.
22. Kiviste A, Sims A, Laarmann D, Hordo M, Lilleleht A, Kangur A, Korjus H: *The Estonian Permanent Forest Plots Network*. Proceedings of an International Workshop at Beijing Forestry University; 2012:97-108.

23. Kramer H: *Waldwachstumslehre*. Hamburg and Berlin: Verlag Paul Parey; 1988:374.

24. Kuehl RO: *Statistical Principles of Research Design and Analysis*. Belmont, California: Duxbury Press; 1994:686.

25. LaBau VJ, Bones JT, Kingsley NP, Lund HG, Smith WB: A History of U.S. Department of Agriculture Forest Service Forest Survey, 1830–2004. In *Proceedings of the Seventh Annual Forest Inventory and Analysis Symposium; October 3–6, 2005; Portland, ME. Gen. Tech. Rep. WO-77*. Edited by McRoberts RE, Reams GA, Van Deusen PC, McWilliams WH. Washington, DC: U.S. Department of Agriculture, Forest Service; 2007:9-10.

26. Laiho O, Pukkala T, Lähde E: Height increment of understorey Norway spruces under different tree canopies. *For Ecosyst* 2014, 1:4.

27. Lee WK: *Wachstums- und Ertragsmodelle für Pinus Densiflora in der Kangwon-Provinz, Korea. Diss., Forstwiss*. Fachbereich, Univ. Göttingen: Cuvillier Verlag; 1993:178 S.

28. Lessard VC, Van Deusen Paul C: Updating Indiana Annual Forest Inventory and Analysis Plot Data Using Eastern Broadleaf Forest Diameter Growth Models. In*Proceedings of the Second Annual Forest Inventory and Analysis Symposium; 2000 October 17–18; Salt Lake City, UT. Gen. Tech. Rep. SRS-47*. Edited by Reams Gregory A, McRoberts Ronald E. Asheville, NC: U.S. Department of Agriculture, Forest Service, Southern Research Station; 2001:66-69.

29. Liu WG, Fan XH, Wang JS, Lu WM, Zhang CY, Gadow K: The effect of 744 above- and belowground competition on the spectral reflectance of Fraxinus 745 mandshurica leaves. *Int J Remote Sens* 2011.in print

30. Longuetaud F, Seifert T, Leban J-M, Pretzsch H: Long-term dynamics of crowns in sessile oak stands by means of spatial analysis. *For Ecol Manag* 2008, 255:2007-2019.

31. Maltamo M, Næsset E, Vauhkonen J (Eds): In *Forestry Applications of Airborne Laser Scanning - Concepts and Case Studies*. 2014, 412. [*Springer Book Series Managing Forest Ecosystems, Vol. 27*]

32. Mason B, Kerr G, Pommerening A, Edwards C, Hale S, Ireland D, Moore R: Continuous Cover Forestry in British Conifer Forests. In

Forest Research (2005). Annual report and accounts 2003–2004; 2005:38-53.

33. Ministerial Conference on the Protection of Forests in Europe: Ministerial conference on the protection of forests in Europe. 2008. http://www.mcpfe.org/
34. Monserud RA: Large-scale management experiments in the moist maritime forests of the Pacific Northwest. *Landsc Urban Plan* 2002, 59(3):159-180.
35. Montréal Process: *Criteria and Indicators for the Conservation and Sustainable Management of Temperate and Boreal Forests*. Montreal: The Montreal Process; 1998. http://www.rinya.maff.go.jp/mpci/
36. Nothdurft A, Schmidt M: *Kalibrierfähige Modelle für Simultane Prognosen von Durchmesser und Höhe bei Waldinventuren – Integrierte Nutzung von Langzeitbeobachtungen auf Versuchsflächen*. 2010, 125-130. [*DVFFA – Sektion Ertragskunde, Jahrestagung 2010*]
37. O'Hehir J: *Growth and Yield Models for South Australian Radiata Pine Plantations Incorporating Fertilising and Thinning*. Institute of Land and Food Resources, University of Melbourne: PhD dissertation; 2001:167.
38. Pelz DR, Kohnle U: Forest Observational Studies in Germany. In *Forest Observational Studies*. Edited by Zhao XH, Zhang CY, Gadow K. Proceedings of an International Workshop at Beijing Forestry University; 2012:109-124.
39. Pretzsch H: *Forest Dynamics, Growth and Yield*. Springer; 2009:684.
40. Pukkala T: Stand management optimization – the role of simplifications. *For Ecosyst* 2014, 1:3.
41. Radtke PJ, Amateis RL, Prisley SP, Copenheaver CA, Chojnacky DC, Püttman JR, Burkhart HE: Modeling production and decay of coarse woody debris in loblolly pine plantations. *For Ecol Manage* 2009, 257:790-799.
42. Seidel D: 3D-Terrestrial laser scanning. 2014. http://wiki.awf.forst.uni-goettingen.de/wiki/index.php/ 3D-_Terrestrial_laser_scanning (date read: 3rd February, 2014)

43. Seymour RS, Guldin J, Marshall DM, Palik B: Large-scale, long-term silvicultural experiments in the United States: historical overview and contemporary examples. *Allg Forst u J Ztg* 2006, 177(6/7):104-112.

44. Seymour RS, Guldin J, Marshall D, Palik B: Large-scale, long-term silvicultural experiments in the United States: historical overview and contemporary examples. *Allg Forst u J Ztg* 2006, 177(6/7):104-112.

45. Sims A, Kangur A, Hordo M, Kiviste A, Jõgiste K, Gadow K: Tracking disturbance-induced changes in stand development at irregular measurement intervals in the järvselja forest experiments. *Balt For* 2009, 15(2):151-160.

46. Spellmann H, Wagner S, Nagel J, Guericke M, Griese F: In der Tradition stehend, neue Wege beschreitend. *Forst u Holz* 1996, 51(11):363-368.

47. Szaro RC, Peterson CE, Gadow K: Operational experiments for sustainably managing forests. *Allg Forst u J Ztg* 2006, 177(6/7):98-104.

48. Temesgen H, Zhang C, Zhao X: Modelling tree height-diameter relationships of multi-species and multi-layered forests: a large observational study from Northeast China. *For Ecol Manag* 2014, 276:139-145.

49. Tewari VP, Sukumar R, Kumar R, Gadow K: Forest observational studies in India - past developments and future prospects. *For Ecol Manage Spec Issue For Observational Stud* 2014.in press

50. Tomppo E, Gschwantner T, Lawrence M, McRoberts RE (Eds): In *National Forest Inventories – Pathways for Common Reporting*. 2010, XXVI:612. [*Springer Book Series Managing Forest Ecosystems*]

51. Valkonen S, Eerikäinen K, Saksa T: Ingrowth, survival and height growth of small trees in uneven-aged Picea abies stands in southern Finland. *For Ecosyst* 2014, 1:5.

52. Vanclay JK: *Modelling Forest Growth - Applications to Mixed Tropical Forests*. Wallingford, UK: CAB International; 1994.

53. Vanclay JK: Unsuspected implications arising from assumptions in simulations: Insights from recasting a forest growth model in system dynamics. *For Ecosyst* 2014, 1:7.

54. Wang J, Zhang C, Xia F, Zhao X, Wu L, Gadow K: Biomass structure and allometry of*Abies nephrolepis* (Maxim) in Northeast China. *Silva Fennica* 2011, 45(2):211-226.

55. Wehenkel C, Corral-Rivas JJ, Hernández-Díaz JC, Gadow K: Estimating balanced structure areas in multi-species forests on the Sierra Madre Occidental, Mexico. *Ann For Sci* 2011, 68:385-394.

56. Zhang CY, Zhao XH, Liu XD, Gadow K: Spatial distributions and spatial associations of dominant tree species in Korean pine broadleaved old-growth forests in Changbai Mountains. *Balt For* 2009a, 16(1):66-74.

57. Zhang CY, Zhao XH, Gadow K: Gender, neighboring competition and habitat effects on the stem growth of dioecious *Fraxinus mandshurica* trees in a northern temperate forest. *Ann For Sci* 2009b, 66:812-821.

58. Zhang CY, Zhao XH, Zhang CY, Zhao XH, Gao L, Gadow K: Gender-related distributions of *Fraxinus mandshurica* in secondary and old-growth forests. *Acta Oecol* 2010a, 36:55-62.

59. Zhang CY, Zhao XH, Gadow K: Partitioning temperate plant community structure at different scales. *Acta Oecol* 2010b, 36:306-313.

60. Zhang CY, Wang J, Zhao XH, Xia FC, Gadow K: Sexual dimorphism in reproductive and vegetative allometry for two dioecious *Rhamnus* plants in north-eastern China. *Eur J For Res* 2012, 131:1287-1296.

61. Zhao XH, Zhang CY, Gadow K: *Forest Observational Networks. Proceedings of the IUFRO-EFI-ICFFI Conference "Ecosystem Design for Multiple Services-with an Emphasis on Eurasian Boreal Forests"*. St. Petersburg/Russia: St. Petersburg Forest Technical University; 2011:34-47. 9–11 November, 2011

62. Zhao XH, Zhang CY, Gadow K: *A Forest Observational Network for China*. Proceedings of an International Workshop at Beijing Forestry University; 2012:1-16.

Chapter 7

Incorporating Novelty and Novel Ecosystems into Restoration Planning and Practice in the 21st Century

Michael P Perring, Rachel J Standish, and Richard J Hobbs

Ecosystem Restoration and Intervention Ecology Research Group, School of Plant Biology, the University of Western Australia, 35 Stirling Highway, Crawley, WA 6009, Australia

ABSTRACT

Novelty pervades the biosphere. In some cases, potentially irreversible abiotic and/or biotic changes have led to the crossing of thresholds and thus the formation of "novel ecosystems." Their widespread emergence (particularly on land) and the presence of continued environmental change challenge a traditional restoration goal of restoring an historical ecosystem. Instead, we argue that restoration could broaden its frame of reference to consider how novel ecosystems might be used to maintain global biodiversity and provide ecosystem services and, in doing so, save potentially wasted efforts in attempting to fulfil traditional goals. Here we explore this contention in more depth by addressing: *Are novel ecosystems innovative planning or lowering the bar?* We show that novel ecosystems were not innovative planning in their original conception. On the contrary, they were recognized as ecosystems that were recalcitrant to traditional restoration approaches, coupled with an awareness that they had arisen inadvertently through deliberate human activity, either on- or off-site. Their recalcitrance to traditional restoration suggests that alternative goals may exist for these ecosystems using sometimes innovative intervention. This management may include biodiversity conservation or restoration for ecological function. We elucidate the latter aspect with reference to an experiment in the wheatbelt of Western Australia—The Ridgefield Multiple Ecosystem Services Experiment—the design of which has been informed by ecological theory and the acceptance of novelty as an ecosystem component. Although novel ecosystems do provide opportunities to broaden restoration planning and practice, and ultimately maintain and conserve global biodiversity in this era of environmental change, they necessarily "lower the bar" in restoration if the bar is considered to be the historical ecosystem. However, in these times of flux, such a bar is increasingly untenable. Instead, careful and appropriate interventions are required at local, regional, and global scales. These interventions need to take history into account, use ecological and evolutionary theory to inform their design, and be mindful of valid concerns such as hubris. Careful interventions thus provide an opportunity for broadening restoration's framework to focus on maintaining global biodiversity and delivering ecosystem services as well as the traditional goals of restoring historical ecosystems.

INTRODUCTION

This Special Issue arose from a symposium at the Society for Ecological Restoration Australasia's inaugural conference, where the following question was posed:

"Are novel ecosystems innovative planning or lowering the bar for restoration and rehabilitation?"

We will explore this question by examining what we understand by the terms "novel ecosystem" and "novelty" and by arguing that novel ecosystems (as originally conceived) are not innovative planning. However, given their rising global prominence, we think they are a necessary consideration for restoration planning and practice in the 21st century. Such consideration exists in the broader context of what goals are appropriate for ecological restoration and for whom, in an ever-changing environment. Notably, the pervasiveness of environmental change questions the utility of traditional restoration goals, yet modifying goals can be perceived as "lowering the bar." At the same time, retaining strict fidelity to historical species composition as the basis for restoration goals in a time of environmental flux overlooks the chance that novel ecosystems may provide opportunities for biodiversity conservation and ecosystem service provision—not to mention potentially wastes scarce resources. In this era of novelty we discuss how other goals for restoration could be implemented for novel ecosystems. The management of novel ecosystems may be perceived as innovative given its departure from the way restoration has traditionally been approached and practiced. We conclude by suggesting firstly that novel ecosystems provide opportunities to broaden the framework of ecological restoration and contribute to biodiversity conservation in an era of global change, and secondly that the question of *where* and *when* to implement these opportunities needs careful con-sideration.

REVIEW

Novelty and Novel Ecosystems

The global environment is changing as a result of the exponential increase in human activity since the start of the industrial era (Vitousek et al. 1997ab; Ellis 2011). Species have responded idiosyncratically to these socio-environmental changes (e.g., Lee and Caporn 1998; Tylianakis et al. 2008) by assembling and reassembling into communities. Given the pervasiveness of change, all ecosystems, whether on land or at sea, possess greater or lesser degrees of novelty, having no strict analogue in present or historic records (Halpern et al. 2008; Williams and Jackson 2007). Novelty per se does not limit the opportunities for restoration and conservation as traditionally practiced (Figure 1). In some instances, however, the scale of change has led to ecosystem states that are arguably not able to be returned, at least practically, to a desired former state due to the crossing of ecological and/or socio-economic thresholds (Hobbs et al. 2009; Hallett et al. 2013). Ecosystems that have been identified as going beyond one or more of these alleged thresholds—the position and presence of which may differ depending on perspective—have been referred to as novel (Hobbs et al. 2006). Although the immediate actions of people were and continue to be deliberate, the consequences of such actions were not foreseen (or, at worst, lessons from previous development activities were deliberately ignored). Subsequently, novel ecosystems can be considered inadvertent consequences of deliberate human actions. Further, novel ecosystems exist both on- and off-site from the anthropogenic source of disturbance (e.g., abandoned agricultural fields vs. ecosystems with elevated nitrogen deposition) (Mascaro et al. 2013). As such, and in their original conception, we would argue that novel ecosystems are not innovative planning. They simply *are*: they do not arise through deliberate planning, innovative or otherwise. Instead, they can be summarized as being distinct from other ecosystems through having a difference in species composition, structure, or function. Secondly, they possess thresholds in these attributes at which change is irreversible, a potentially controversial designation. Lastly, novel ecosystems persist (i.e., are self-organizing) in the absence of continued human intervention (Harris et al. 2013) and have departed from their expected, historical trajectory. These characterizations provide the following working definition for novel ecosystems:

Novel ecosystems are a system of abiotic, biotic and social components (and their interactions) that, by virtue of human influence, differ from those that prevailed historically, having a tendency to self-organize and manifest novel qualities without intensive human management. Novel ecosystems are distinguished from hybrid ecosystems by practical limitations (a combination of ecological, environmental and social thresholds) on the recovery of historical qualities (Hobbs et al.2013a).

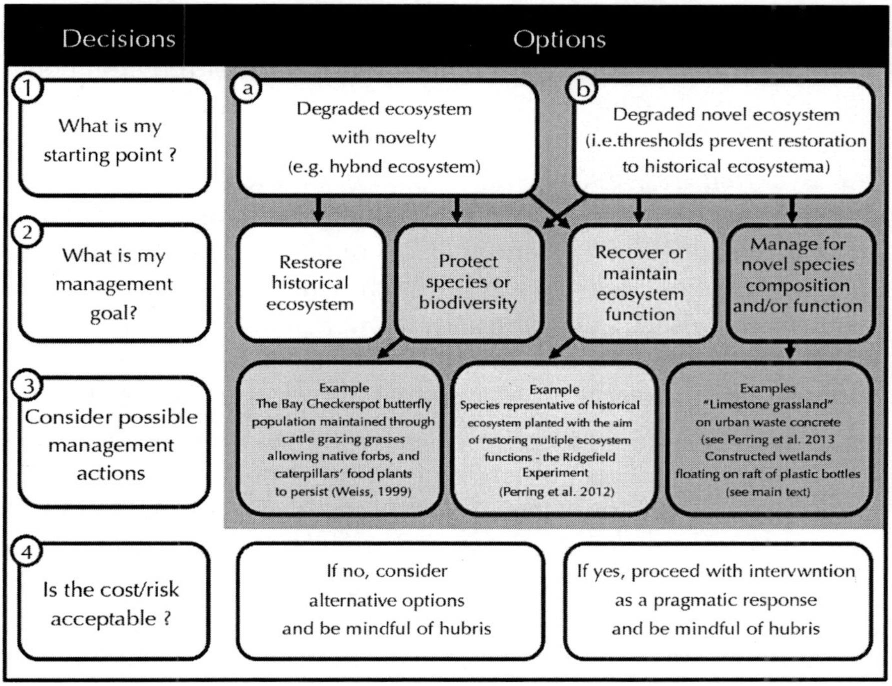

Figure 1: Management options for restoration where (a) novelty and (b) a novel ecosystem have been identified. Decisions need to be taken by working through questions 1–4 (left hand side); depending on your ecosystem's current condition, different goals are options for restoration management as further explored in the main text. Example management actions are provided for ecosystems identified as novel. This figure has been modified from Hulvey et al. 2013.

Whither Restoration gave the Rising Prominence of Novel Ecosystems? The Necessity of Broadening the Restoration Framework

Restoration ecologists have traditionally contemplated moving from an undesired ecosystem state to a desired state, where that desired state is usually one that existed historically (Hobbs and Norton 1996; SERI 2006). This basic (value-laden) tenet of restoration—i.e., that it is desirable to return to the historical reference—is increasingly questioned in an era of multiple, simultaneous, and rapid environmental changes. Indeed, recent meta-analyses have confirmed the sizeable gap between restored and reference ecosystems for projects with the latter as a goal (Rey-Benayas et al. 2009). Furthermore, going back to historic species compositions in a particular area may, *in extremis*, lead to local extinction if environmental conditions change sufficiently to alter conditions beyond species' niches [an "ossification" of the ecosystem (Harris et al. 2006)]. The identification of ecosystems as novel can therefore be useful in avoiding waste of scarce resources in potentially futile attempts to return to the historical reference state.

If novel ecosystems were rare across the biosphere, there may be little reason to consider them in restoration and conservation frameworks more broadly. However the pervasiveness of environmental change and human appropriation of natural resources suggests that this is an unlikely scenario. Indeed, recent global-scale estimates suggest between 35 and 40% of the terrestrial ice-free land surface may constitute novel ecosystems, with one estimate suggesting that novel ecosystems cover more land surface area than unused "wild" landscapes (Perring and Ellis 2013; Ellis et al. 2010).

Although such global-scale estimates of novelty clearly outline the necessity of considering novel ecosystems in restoration and rehabilitation, digging deeper into how estimates were calculated raises questions (Perring and Ellis 2013). Most importantly, estimates do not attempt to differentiate novel ecosystems from those that exhibit facets of novelty but are considered to be able to be returned to a reference state [termed "hybrid" systems by Hobbs et al. (2009) as noted in the definition]. Hobbs et al. (2013a) suggest that hybrid and novel ecosystems differ due to the crossing of thresholds such that abiotic or biotic changes are currently considered irreversible in novel ecosystems, for either ecological or socio-economic reasons. The identification of such thresholds remains an important constraint; arguably, virtually nothing is irreversible given enough input (e.g., Sherkow and Greely 2013). However, the recognition of limitations on restoration because of ecological change or socio-economic constraints is important for planning and setting priorities.

Notwithstanding the issues outlined above, the idea that novelty is pervading the biosphere is uncontroversial even if it is regretted by some. Indeed, it is expected that the novelty of ecosystems will increase over time (Jackson 2013). Given this increasing novelty, it is more likely that thresholds associated with irreversible change will be crossed, and novel ecosystems will therefore increase in prominence. Since novel ecosystems, by definition, resist traditional restoration efforts it would be tempting to ignore them, but the increasing prominence of novelty prompts us to consider what place novel ecosystems might have in restoration planning and practice. If novel ecosystems do have a place in restoration, can we set socially acceptable and innovative goals for their management? If so, are such goals lower than the bar set by traditional ecological restoration?

Managing Novel Ecosystems: Innovation in Restoration Planning and Practice?

Ecological restoration's basic tenets are under question in the context of rapidly changing environments. Restoring to historical ecosystems is unlikely to be a realistic or practical goal for novel ecosystems. We therefore need to consider how to amend the tenets of restoration while not losing sight of the broader goals of biodiversity conservation and the delivery of ecosystem services. We suggest that ecological restoration requires a broadening of its frame of reference in which historical ecosystem composition, and particularly function, inform planning but do not place strictures upon it. In short, we think there is a necessity for restoration policy, planning, and practice to take account of the presence of novel ecosystems. Restoration ecologists could pave the way for innovative management practices by using novel ecosystems as test beds for new ideas (Hobbs et al. 2013b). Ultimately, testing new ideas via research and restoration practice will inform the management of novel ecosystems.

This test-bed concept has been utilized in the design of an experiment in the wheatbelt of south-west Western Australia, where we (the authors) and others are investigating how compositional differences among restored assemblages of native species influence ecosystem function on degraded agricultural land (Perring et al. 2012). Habitat fragmentation and climate change, together with non-native species and high nutrient levels, can prevent the recovery of historical ecosystems (e.g., Prober and Smith 2009; Pywell et al.

2007). Furthermore, old-fields that do not recover without assistance are likely to be resistant to traditional restoration efforts (Cramer et al.2008). Given the presence of multiple, potentially irreversible changes, abandoned agricultural lands are therefore one of the places where novel ecosystems can develop (Hobbs et al. 2006).

At the experimental site, ecological and socio-economic thresholds are present which indicated that restoration to historic York gum woodland would likely be prevented (Standish et al. 2007). Chief among the changes that are difficult to reverse are missing soil seed banks, nutrient enrichment, the presence of competitive non-native species, and reduced seed sources and land degradation associated with habitat fragmentation (Standish et al. 200620072008). Socio-economically, there is not the current knowledge to recover all facets of the historic ecosystem (e.g., native ground cover), and additionally there is not the needed level of investment to discern methods to do so. We have also heard concerns from some farmers about widespread re-vegetation leading to increased fire risk as well as opportunity cost. Whether these changes are truly irreversible is hard to determine, but we assumed that without intervention the experimental site would constitute an unmanaged novel ecosystem now and into the future, particularly given the further context of the drying climate of the region (CSIRO 2007).

Where a novel ecosystem is identified and the restoration of the historic ecosystem is precluded, management options can broaden from those focused on the restoration of historical baselines (Figure 1). Other goals for novel ecosystems include biodiversity conservation (of one or multiple species) and the restoration of ecosystem functions (Low 2003; Hallett et al. 2013; Hulvey et al.2013). In ecology, the meaning of "function" can relate to processes, such as efficient nutrient cycling and storage of carbon, or to the roles that particular species play, or may be applied to the system as a whole in terms of its "health" or otherwise (Jax 2005). Restoration of mined landscapes, for example, often aims to restore the abiotic processes that sustain ecosystem functions through the reconstruction of soils, slope, and landscape hydrology [i.e., rehabilitation (Bradshaw 1997)].

At the experimental site, we set goals focused on the provision of combinations of ecosystem functions and services through the restoration of elements of the historical ecosystem. We defined functions in terms of ecosystem processes and concentrated on how we might design restoration plantings to support such processes, particularly those valued by people, i.e., ecosystem services (Millennium Ecosystem Assessment 2005). Thus we developed the Ridgefield Multiple Ecosystem Services Experiment, which will include measurement of these services: carbon sequestration, invasion resistance to weeds, and reduction of excess soil nutrients to avoid eutrophication (Perring et al. 2012).

The Ridgefield Multiple Ecosystem Services Experiment: A Test Bed for Restoration of Ecosystem Functions in Novel Ecosystems

We chose to plant native vegetation with different traits which we hypothesized would influence the provision of our targeted ecosystem services based on our knowledge of ecological theory (e.g., Hooper et al. 2005; Cardinale et al. 2012; McGill et al. 2006). Functional groups are defined by rooting depth at maturity and nutrient acquisition traits including whether plants form cluster-roots, fix nitrogen, and/or associate with arbuscular and ecto-mycorrhizal fungi (see also Table two in Perring et al. 2012). In addition to a gradient of species and functional richness, two experimentally manipulated environmental changes increase abiotic and biotic novelty. Nitrogen deposition is being simulated by the application of dry ammonium nitrate to mimic the projected maximum rate of nitrogen deposition in this area around 2050 (Phoenix et al. 2006). Non-native herbaceous species are being removed through the application of broad-spectrum herbicide (glyphosate) at least twice per year. With these manipulations we are asking whether the presence of a novel component (non-native weed species) in different environmental contexts (with and without nitrogen deposition) affects the delivery of ecosystem services. The restored systems also possess novelty on at least three counts compared with the reference York gum woodland: they are planted at greater density, lack a forb-rich native understory (Prober et al. 2011), and are (essentially) even-aged (Figure 2).

Figure 2: Ridgefield before (2 August 2010) and after (17 April 2012) planting to restore multiple functions. Photo credits: M.P. Perring (left) and Nancy A. Shackelford (right).

The managed novel ecosystem at Ridgefield may have a number of benefits consistent with the goals of traditional restoration. For instance, plots with greater plant biodiversity and more structural complexity will likely support more species than a simple mono-culture [e.g., birds (Munro et al. 2011)]. In addition, the planting forms a block of vegetation that links remnant patches of native vegetation. Thus available habitat has increased, which may be of particular importance to wildlife that requires large areas of habitat for foraging and survival. However, water relations could be negatively affected, which in turn may have indirect effects on biodiversity (Bradshaw et al. 2013). On the whole, we argue that deliberate intervention after agricultural abandonment, as outlined by reference to Ridgefield, is an innovative response towards ensuring the ongoing supply of ecosystem services and biodiversity conservation within the agricultural landscape. The ecosystem remains novel due to its component parts and underlying environmental changes (e.g., high soil nutrient levels), but it has historical echoes through the use of native canopy species that we chose based on an expectation that they would be somewhat adapted to the environmental conditions of the site. Ridgefield has become a test bed for exploring how to manage novel ecosystems that have emerged from abandoned agricultural land in a way that acknowledges the environmental changes of the 21st century. This test-bed concept could be extended to include how genetic provenance also influences the provision of ecosystem services in a changing environment (e.g., Whitham et al. 2008).

Broadening Restoration Goals through Novel Ecosystems

Deliberate creation of novel ecosystems, sometimes without historical precedent, may also have a place in achieving restoration goals more broadly, although context is important. For instance, urban ecosystems, remnants or otherwise, have vastly changed abiotic and biotic conditions compared to the ecosystems that occupied the landscape before urbanization (McDonnell and Pickett 1990; Paul and Meyer 2001; Pickett et al. 2001). Ongoing fragmentation and development superimposed on this changed palette make a restoration goal of returning to the historical reference somewhat moot (Perring et al. 2013). Instead, a pragmatic approach might be to manage urban ecosystems for multiple benefits. For example, floating wetlands made from rafts of plastic bottles have improved habitat quality in polluted waterways in Baltimore and elsewhere on the eastern seaboard of the United States. Not only has this project had water quality and habitat benefits, but it has also addressed social ills and made local youth in the area feel empowered (Keith Bowers 2012, pers. comm.).

We stress here that deliberate *creation* of a novel ecosystem as described above is, we think, a potentially pragmatic and innovative solution for the management of novel conditions that have arisen inadvertently. However, we do not mean to suggest that novel ecosystems be created in places where there is the potential to achieve goals akin to traditional restoration and conservation practices. Indeed, while the creation of novel ecosystems may be a sensible approach to ecosystem management in some cases, ecologists, restoration practitioners, and planners must be mindful of hubris or overestimating the ability to understand and manage complex ecosystems (Standish et al. 2013). These points highlight again the need to establish metrics about when interventions may be sensible or otherwise, depending upon ecosystem state and landscape context (e.g., Kirkpatrick and Kiernan 2006).

Accepting Novel Ecosystems: Lowering the Bar in Restoration Planning and Practice?

There are a number of concerns about the management of novel ecosystems, beyond hubris, that warrant discussion and provide a counterweight to the utilitarian arguments outlined above. Perhaps the most widely held concern is the sense that their incorporation into management frameworks could lead to a lowering of the bar for restoration practice, as made clear by the organizers of the Special Issue. The "bar" in this case is the historical ecosystem, the goal of restoration as traditionally practiced. The setting of the "bar" in reality depends upon the goals set for restoration—some defend the traditional notion of restoration and wish to restrict the term to activities that fit a relatively limited definition (e.g., Clewell and Aronson 2013), whereas others have argued that the traditional perspective is too narrow to encompass all the types of activity undertaken to achieve various types of environmental repair (Hobbs 2007; Suding 2011). Different goals will have different "bars." Hence the discussion of novel ecosystems is part of a broader debate on what restoration is or should be. If we accept that the historic ecosystem is an unobtainable, possibly unknowable, goal for novel ecosystems, then pragmatically a different goal must be set for their management. Accepting this fact will be difficult for some people, particularly if they are also concerned that the goals for the management of novel ecosystems are somehow less valuable than the historic ecosystem (Light et al. 2013). Such concerns place a premium on the proper diagnosis of a novel ecosystem (i.e., one where change has gone beyond a threshold, making it irreversible), although novelty, in the absence of thresholds, will likely have to be considered as restoration moves into the future. Additionally, the environmental virtues of those people involved in the decision-making process for the management of novel ecosystems are arguably more critical than they are for traditional restoration practice because there is more choice about the goal to set and how to achieve it (Light et al. 2013). The challenge that global environmental change brings to the practice of environmental management is not unique to novel ecosystems. Indeed, there is evidence to suggest that traditional conservation practice is also evolving to meet these challenges, with similar potential for controversy (e.g., assisted migration, cryo-preservation, seed banking).

The concern that people will exercise human ambition in the management of novel ecosystems is a valid one (Standish et al. 2013) and may be a persistent concern in that it is not readily argued away. On the other hand, some concerns about novel ecosystems are misapprehensions, including a belief that invasive species will be accepted, that acceptance of novel ecosystems will replace traditional conservation and restoration practices, and that our Western consumer culture lends itself to accepting novel ecosystems into management frameworks simply because they are "new." These misapprehensions can be argued away convincingly—invasive species, whether native or non-native, should still be tackled where they pose a threat—but measures of impact may provide a pragmatic approach where prevention has already failed (Shackelford et al. 2013). The management framework of novel ecosystems provides a broadening of traditional frameworks, not a replacement. Indeed, it is distinctly possible that history will become more important in restoration practice as it provides the cultural and moral guidance alluded to above (Eric Higgs 2012, pers. comm.). The pervasiveness of consumer culture and its valuation of new can be less readily argued away and so gives us pause to reflect and "places a premium on anticipating these patterns" (Standish et al. 2013).

CONCLUSIONS

Novelty across the biosphere is pervasive. In some cases, human-caused biotic and abiotic changes associated with intensive land use can result in the crossing of thresholds (ecological, economic, and/or social) which prevent the return of historical ecosystems and instead lead to the formation of novel ecosystems. Since environmental changes and human influence are here to stay for the foreseeable future, novel ecosystems, or novelty at the very least, must necessarily be considered in restoration planning and practice. We argue this requires the broadening of traditional conservation and restoration management frameworks to incorporate goals for the management of novel ecosystems. One alternative goal is to restore ecosystem services, and test beds are required to investigate ecologically and ethically sound ways of doing this (e.g., Ridgefield). Other management goals for novel ecosystems include conservation of desired species through the use of novel components or the creation of new ecosystems to achieve desired goals (Figure 1). Recognition of novel ecosystems may save wasted restoration effort and better direct scarce resources to help maintain the globe's unique biodiversity and the ecosystem services that are underpinned by it (Hallett et al. 2013). However, these contentions, and particularly the identification of thresholds beyond which change is irreversible, need to be empirically tested, with options weighed carefully in the balance. Ultimately, humanity needs to find ways to lessen its impact upon ecosystems to help prevent environmental degradation such that our unique global biodiversity, and the evolutionary processes that promote it, can be maintained.

AUTHORS' CONTRIBUTIONS

MPP conceived of and wrote the paper with RJS making an equally substantial contribution. RJH commented on previous drafts and guided the development of the manuscript. All authors read and approved the final version.

ACKNOWLEDGEMENTS

We thank the Australian Research Council for funding, through a Laureate Fellowship to RJH. We also thank the organizers and participants at the Novel Ecosystems special session of the inaugural SERA Conference for comments on MPP's presentation, where ideas featured in this paper were discussed. We especially thank Kris Hulvey for her willingness to share ideas featured in this manuscript and her comments on previous drafts. Finally, we appreciate the diligence and comments from two anonymous reviewers and P. Audet on a previous version of this manuscript.

REFERENCES

1. Bradshaw A (1997) Restoration of mined lands—using natural processes. Ecol Eng 8:255-269
2. Bradshaw CJA, Bowman DMJS, Bond NR, Murphy BP, Moore AD, Fordham DA, Thackway R, Lawes MJ, McCallum H, Gregory

SD, Dalal RC, Boer MM, Lynch AJJ, Bradstock RA, Brook BW, Henry BK, Hunt LP, Fisher DO, Hunter D, Johnson CN, Keith DA, Lefroy EC, Penman TD, Meyer WS, Thomson JR, Thornton CM, VanDerWal J, Williams RJ, Keniger L, Specht A (2013) Brave new green world—consequences of a carbon economy for the conservation of Australian biodiversity. Biol Conserv 161:71-90

3. Cardinale BJ, Duffy JE, Gonzalez A, Hooper DU, Perrings C, Venail P, Narwani A, Mace GM, Tilman D, Wardle DA, Kinzig AP, Daily GC, Loreau M, Grace JB, Larigauderie A, Srivastava DS, Naeem S (2012) Biodiversity loss and its impact on humanity. Nature 486:59-67
4. Clewell AF, Aronson J (2013) Ecological restoration: principles, values and structure of an emerging profession. Washington D.C: Island Press.
5. Cramer VA, Hobbs RJ, Standish RJ (2008) What's new about old fields? Land abandonment and ecosystem assembly. Trends Ecol Evol 23(2):104-112
6. CSIRO (2007) Climate change in Australia: technical report. CSIRO. http://www.climatechangeinaustralia.gov.au/documents/resources/TR_Web_FrontmatterExecSumm.pdf
7. Ellis EC (2011) Anthropogenic transformation of the terrestrial biosphere. Phil Trans R Soc A 369:1010-1035
8. Ellis EC, Klein Goldewijk K, Siebert S, Lightman D, Ramankutty N (2010) Anthropogenic transformation of the biomes, 1700 to 2000. Glob Ecol Biogeogr 19:589-606
9. Hallett LM, Standish RJ, Hulvey KB, Gardener MR, Suding KN, Starzomski BM, Murphy SD, Harris JA (2013) Towards a conceptual framework for novel ecosystems. In: Hobbs RJ, Higgs ES, Hall CM (eds) Novel ecosystems: intervening in the new ecological world order, Chichester: Wiley. pp 16-28
10. Halpern BS, Walbridge S, Selkoe KA, Kappel CV, Micheli F, D'Agrosa C, Bruno JF, Casey KS, Ebert C, Fox HE, Fujita R, Heinemann D, Lenihan HS, Madin EMP, Perry MT, Selig ER, Spalding M, Steneck R, Watson R (2008) A global map of human impact on marine ecosystems. Science 319:948-952
11. Harris JA, Hobbs RJ, Higgs E, Aronson J (2006) Ecological restoration and global climate change. Restor Ecol 14:170-176

12. Harris JA, Murphy SD, Nelson CR, Perring MP, Tognetti PM (2013) Characterizing novel ecosystems: challenges for measurement. In: Hobbs RJ, Higgs ES, Hall CM (eds) Novel ecosystems: intervening in the new ecological world order, Chichester: Wiley. pp 192-204
13. Hobbs RJ (2007) Setting effective and realistic restoration goals: key directions for research. Restor Ecol 15(2):354-357
14. Hobbs RJ, Norton DA (1996) towards a conceptual framework for restoration ecology. Restor Ecol 4(2):93-110
15. Hobbs RJ, Arico S, Aronson J, Baron JS, Bridgewater P, Cramer VA, Epstein PR, Ewel JJ, Klink CA, Lugo AE, Norton D, Ojima D, Richardson DM, Sanderson EW, Valladares F, Vila M, Zamora R, Zobel M (2006) Novel ecosystems: theoretical and management aspects of the new ecological world order. Glob Ecol Biogeogr 15:1-7
16. Hobbs RJ, Higgs E, Harris JA (2009) Novel ecosystems: implications for conservation and restoration. Trends Ecol Evol 24(11):599-605
17. Hobbs RJ, Higgs ES, Hall CM (2013) Defining novel ecosystems. In: Hobbs RJ, Higgs ES, Hall CM (eds) Novel ecosystems: intervening in the new ecological world order, Chichester: Wiley. pp 58-60
18. Hobbs RJ, Higgs ES, Hall CM (2013) What do we know about, and what do we do about, novel ecosystems? In: Hobbs RJ, Higgs ES, Hall CM (eds) Novel ecosystems: intervening in the new ecological world order, Chichester: Wiley.
19. Hooper DU, Chapin FS, Ewel JJ, Hector A, Inchausti P, Lavorel S, Lawton JH, Lodge DM, Loreau M, Naeem S, Schmid B, Setala H, Symstad AJ, Vandermeer J, Wardle DA (2005) Effects of biodiversity on ecosystem functioning: a consensus of current knowledge. Ecol Monogr 75(1):3-35
20. Hulvey KB, Standish RJ, Hallett LM, Starzomski BM, Murphy SD, Nelson CR, Gardener MR, Kennedy PL, Seastedt TR, Suding KN (2013) Incorporating novel ecosystems into management frameworks. In: Hobbs RJ, Higgs ES, Hall CM (eds) Novel ecosystems: intervening in the new ecological world order, Chichester: Wiley.

21. Jackson ST (2013) Perspective: ecological novelty is not new. In: Hobbs RJ, Higgs ES, Hall CM (eds) Novel ecosystems: intervening in the new ecological world order, Chichester: Wiley. pp 63-65
22. Jax K (2005) Function and "functioning" in ecology: what does it mean? Oikos 111(3):641-648
23. Kirkpatrick J, Kiernan K (2006) Natural heritage management. In: Lockwood M, Worboys GL, Kothari A (eds) Managing protected areas: a global guide, London: Earthscan. pp 421-447
24. Lee JA, Caporn SJM (1998) Ecological effects of atmospheric nitrogen deposition on semi-natural terrestrial ecosystems. New Phytol 139:127-134 Publisher Full Text
25. Light A, Thompson A, Higgs ES (2013) Valuing novel ecosystems. In: Hobbs RJ, Higgs ES, Hall CM (eds) Novel ecosystems: intervening in the new ecological world order, Chichester: Wiley.
26. Low T (2003) the new nature. London: Penguin.
27. Mascaro J, Harris JA, Lach L, Thompson A, Perring MP, Richardson DM, Ellis E (2013) Origins of the novel ecosystems concept. In: Hobbs RJ, Higgs ES, Hall CM (eds) Novel ecosystems: intervening in the new ecological world order, Chichester: Wiley. pp 45-57
28. McDonnell MJ, Pickett STA (1990) Ecosystem structure and function along urban–rural gradients: an unexploited opportunity for ecology. Ecology 71(4):1232-1237
29. McGill BJ, Enquist BJ, Weiher E, Westoby M (2006) Rebuilding community ecology from functional traits. Trends Ecol Evol 21(4):178-185
30. Millennium Ecosystem Assessment (2005) Ecosystems and human well-being: biodiversity synthesis. Washington D.C: World Resources Institute.
31. Munro NT, Fischer J, Barrett G, Wood J, Leavesley A, Lindenmayer DB (2011) Bird's response to revegetation of different structure and floristics—are "restoration plantings" restoring bird communities? Restor Ecol 19(201):223-235
32. Paul MJ, Meyer JL (2001) Streams in the urban landscape. Annu Rev Ecol Evol Syst 32:333-365
33. Perring MP, Ellis EC (2013) The extent of novel ecosystems: long in time and broad in space. In: Hobbs RJ, Higgs ES, Hall CM

(eds) Novel ecosystems: intervening in the new ecological world order, Chichester: Wiley. pp 66-80

34. Perring MP, Standish RJ, Hulvey KB, Lach L, Morald TK, Parsons R, Didham RK, Hobbs RJ (2012) The Ridgefield Multiple Ecosystem Services Experiment: can restoration of former agricultural land achieve multiple outcomes? Agric Ecosyst Environ 163:14-27
35. Perring MP, Manning P, Hobbs RJ, Lugo AE, Ramalho C, Standish RJ (2013) Novel urban ecosystems and ecosystem services. In: Hobbs RJ, Higgs ES, Hall CM (eds) Novel ecosystems: intervening in the new ecological world order, Chichester: Wiley.
36. Phoenix GK, Hicks WK, Cinderby S, Kuylenstierna JCI, Stock WD, Dentener FJ, Giller KE, Austin AT, Lefroy RDB, Gimeno BS, Ashmore MR, Ineson P (2006) Atmospheric nitrogen deposition in world biodiversity hotspots: the need for a greater global perspective in assessing N deposition impacts. Glob Chang Biol 12:470-476
37. Pickett STA, Cadenasso ML, Grove JM, Nilon CH, Pouyat RV, Zipperer WC, Costanza R (2001) Urban ecological systems: linking terrestrial ecological, physical, and socioeconomic components of metropolitan areas. Annu Rev Ecol Evol Syst 32:127-157
38. Prober SM, Smith FP (2009) Enhancing biodiversity persistence in intensively used agricultural landscapes: a synthesis of 30 years of research in the Western Australian wheatbelt. Agric Ecosyst Environ 132:173-191
39. Prober SM, Standish RJ, Wiehl G (2011) After the fence: vegetation and topsoil condition in grazed, fenced and benchmark eucalypt woodlands of fragmented agricultural landscapes. Aust J Bot 59:369-381
40. Pywell RF, Bullock JM, Tallowin JB, Walker KJ, Warman EA, Masters G (2007) Enhancing diversity of species poor grasslands: an experimental assessment of multiple constraints. J Appl Ecol 44:81-94
41. Rey-Benayas JM, Newton AC, Diaz A, Bullock JM (2009) Enhancement of biodiversity and ecosystem services by ecological restoration: a meta-analysis. Science 325:1121-1124

42. SERI (2006) The SER international primer on ecological restoration. Tucson, AZ: SERI.
43. Shackelford N, Hobbs RJ, Heller NE, Hallett LM, Seastedt TR (2013) Finding a middle-ground: the native / non-native debate. Biol Conserv 158:55-62
44. Sherkow JS, Greely HT (2013) What if extinction is not forever? Science 340:32-33
45. Standish RJ, Cramer VA, Hobbs RJ, Kobryn HT (2006) Legacy of land-use evident in soils of Western Australia's wheatbelt. Plant Soil 280:189-207
46. Standish RJ, Cramer VA, Wild SL, Hobbs RJ (2007) Seed dispersal and recruitment limitation are barriers to native recolonization of old-fields in western Australia. J Appl Ecol 44:435-445
47. Standish RJ, Cramer VA, Hobbs RJ (2008) Land-use legacy and the persistence of invasive *Avena barbata* on abandoned farmland. J Appl Ecol 45:1576-1583
48. Standish RJ, Thompson A, Higgs ES, Murphy SD (2013) Concerns about novel ecosystems. In: Hobbs RJ, Higgs ES, Hall CM (eds) Novel ecosystems: intervening in the new ecological world order, Chichester: Wiley.
49. Suding KN (2011) Towards an era of restoration in ecology: successes, failures, and opportunities ahead. Annu Rev Ecol Evol Syst 42:465-487
50. Tylianakis JM, Didham RK, Bascompte J, Wardle DA (2008) Global change and species interactions in terrestrial ecosystems. Ecol Lett 11:1351-1363
51. Vitousek PM, Aber JD, Howarth RW, Likens GE, Matson PA, Schindler DW, Schlesinger WH, Tilman DG (1997) Human alteration of the global nitrogen cycle: sources and consequences. Ecol Appl 7(3):737-750
52. Vitousek PM, Mooney HA, Lubchenco J, Melillo JM (1997) Human domination of Earth's ecosystems. Science 277:494-499
53. Whitham TG, DiFazio SP, Schweitzer JA, Shuster SM, Allan GJ, Bailey JK, Woolbright SA (2008) Extending genomics to natural communities and ecosystems. Science 320:492-495

54. Williams JW, Jackson ST (2007) Novel climates, no-analog communities, and ecological surprises. Front Ecol Environ 5(9):475-482

Chapter 8

Development of a Fuel Efficient Cookstove through a Participatory Bottom-Up Approach

Vijay H Honkalaskar[1], Upendra V Bhandarkar[2], and Milind Sohoni[3]

[1]CTARA, Indian Institute of Technology Bombay, Mumbai, 400076, India

[2]Department of Mechanical Engineering, Indian Institute of Technology Bombay, Powai, Mumbai, 400076, India

[3]Department of Computer Science and Engineering, Indian Institute of Technology Bombay, Mumbai, 400076, India

ABSTRACT

Background

Since 1940s, other than a few success stories, the outcomes of efforts of development and dissemination of improved cookstoves[a] have not been so fruitful. This paper presents a bottom-up approach that was successfully implemented to develop a fuel-efficient cookstove in a tribal village that has resulted in a substantial reduction in firewood consumption.

[a]Improved cookstove is generally regarded as a relative concept. Improved cookstoves may improve fuel efficiency, reduce indoor air pollution that causes multiple health hazards,

Method

The approach ensured people's participation at multiple stages of the process that started from project selection by capturing people's needs/desires and studying the existing cooking practice to understand its importance in the local context. The performance of the cookstoves was evaluated by modifying a standard Water Boiling Test to accommodate the existing cooking practice. The improvement of the cookstove was achieved by fabricating a simple twisted tape assembly that could be placed on it without changing the existing cookstove.

Results

The optimization of the twisted tape device was first carried out in the laboratory and then implemented in the field. The field-level tests resulted in reduction of firewood consumption by around 21% which is a substantial improvement for such a device. It was also found that the improvement reduced soot [b] accumulation by around 38% and time of cooking preparations by around 18.5%.

[b]It is a particulate matter especially black carbon.

Conclusion

Overall, a bottom-up and participatory process that not only addressed people's perceived needs but also ensured no changes in the existing cooking practice while providing an easy, low cost (around US$1.25) [c], and locally manufacturable solution led to a highly successful improvement in the local cookstove that was accepted easily by the villagers.

[c]It was assumed that US$1 is around 50 Indian Rupees.

BACKGROUND

The inception of a development discourse in the West around the 1940s [1] has led to development of new technologies to address the problems of the poor in undeveloped and developing countries. Since then, the development and dissemination of improved cookstoves has been one of the key areas for development practitioners in government, volunteer organizations, and research laboratories [2,3]. Other than a few success stories, the outcomes of efforts of these practitioners have not been so fruitful. This section summarizes impacts and challenges of different improved cookstove programs over the past 50 years. Sesan [4] broadly categorized these programs in three phases. We have adopted this categorization in the present context of technology development. These three phases are enumerated below.

Early Programs (1950 to 1980)

Since the 1950s, the development and dissemination of improved cookstoves in the developing countries [3] was, to a large extent, due to initiatives from western countries. According to Crewe[5], fashionable aims such as 'working with poor women', 'raising the income of poor artisans', and 'conserving environmental resources' promoted the involvement of international organizations in the development of improved cookstoves. The main goal of this phase for improved cookstoves was to increase the thermal efficiency of the cooking process to reduce the envisaged energy gap [5] between the supply and demand of biomass fuel due to the thinning of the forests. These early programs assumed that people would adopt the improved stoves quickly and that an initial intervention would lead to a self-sustaining program. It was also assumed that there existed only a need to distribute improved cookstoves, and since these were intrinsically superior to the traditional cookstoves, a decrease in fuel consumption by 75% or even more was imminent [6]. As a consequence of these assumptions, the attempts in this phase failed. The clearest of the lessons learned from early experience is that the chances of success are enhanced when people have an explicit need to save fuel, when the new stoves are a significant improvement over the local traditional stoves, and when stoves can be made readily available by local industries or artisans at affordable prices.

Context Responsive Phase (1980 to 1990)

Researchers started to understand that there is a weak link between deforestation and traditional cookstoves as people cut only small branches to fire their cookstoves [6]. The main objective of the cookstove programs was changed to improve the efficiency of the cookstove to address the perceived needs of users to reduce drudgery involved in the firewood fetching activity. Reduction in indoor air pollution was a secondary objective. Researchers started to emphasize the involvement of local people in the cookstove programs. Yet, these programs faced significant problems in gaining the acceptance of the rural poor [6]. There have been a few partial success stories, mainly in urban areas where fuel has to be bought, and hence, cookstove programs became economical. These include 'Jiko' and 'Upesi' stoves in Kenya [7], Chinese national stoves program [8], 'Anagi' stoves in Sri Lanka, and 'Sara-ole' stove in Karnataka, India [9]. There were two nationwide programs implemented by India and China that followed two different dissemination routes. The Indian National Program on Improved Cookstoves (NPIC) was started in 1983. More than 2.8 million stoves were disseminated until the year 2002 when the government of India officially withdrew funding support from the NPIC. The dissemination of the improved cookstoves was supported by government subsidies. Although it had a high dissemination number, NPIC is not recognized as a successful program because of a very low acceptance rate by users [10-12]. The Chinese National Improved Stoves Programme (NISP), which followed a market-based approach, disseminated 129 million stoves from 1982 to 1992 covering 65% of the Chinese rural population. Most of the cost of stove material and construction labor was paid by the users. However, it must be noted that during this period, NISP operated in relatively middle income areas, and its benefit did not reach the rural poor at that time [8]. The more self-sustaining market-based approach adopted by NISP has led the international government and non-government agencies to follow a market-based approach in the third phase [13].

Third Phase (1990 Onwards): Market-Based Approaches

The focus of improved cookstove programs for this phase has been to reduce indoor air pollution along with increasing cookstove efficiency [4]. Indoor air pollution caused by smoke from traditional cookstoves is responsible for nearly 4.3% of the total global diseases and causes more than 1.6 million deaths each year [14]. Drawing on the experience of the second phase, the market approach was emphasized [13]. For example, in India, there was an emergence of multiple stove designs and dissemination programs funded by private agencies. These involved Shell foundation, Envirofit International, Prakti Design, Selco India, B.P., and TIDE [15,16]. These have resulted in limited penetration, owing to high initial cost and low flexibility towards the use of local fuels [11,16]. Learning from these recent initiatives and outcomes of NPIC, the Indian government started the National Biomass Cookstove Initiative in 2009 [11,17], with the aim of providing energy service comparable to clean sources such as LPG while using the same solid biomass fuel with the combined objectives of fuel efficiency, health protection, and climate impacts. To ensure good quality control and reliable performance, this initiative lays an emphasis on centralized manufacturing facilities. There is narrower tolerance to biomass size and moisture content. Thus, there is an additional requirement of fuel processing at household level.

The Global Alliance for Clean Cookstoves was initiated in year 2010 [17]. Coordinated through United Nations Foundation, it has a goal of providing 100 million clean cookstoves by year 2020 which would reduce the health and other impacts of current cooking practices. In spite of all these efforts, what still remains is to make biomass stoves that are not only clean enough to have major health benefits but also affordable by the poor population of the world.

It can be noticed that most of the approaches adopted for improved cookstoves have remained top down and focused essentially on tangible technical objectives. Many researchers such as Bailis et al. [13] and Barnes et al. [6] acknowledge that although the market-based approach can achieve a sustainable dissemination model, it may filter out the poorer section of rural community. Nevertheless, most of these reports do not analyze the development and dissemination approach in the wider social, cultural, ecological, and economical context in which rural people construct their livelihoods.

Learning from the Improved Cookstove Programs

Goals and strategies to develop and disseminate improved cookstoves have changed over the three phases. These were based on the perceived needs/problems and perceived socio-economic context of community members. There have been a few success stories, where not only have the perceived needs and priorities of the community matched with reality, but this identification has been also followed by sustainable dissemination. Analysis of experiences of a number of (seven) improved cookstove programs around the world revealed the following facts:

1. Community Participation: It has been widely appreciated that people's participation is very crucial for the success of any development project. After the failure of the first round of improved cookstove projects until 1980, development agencies sought to adopt more participatory ways to develop and disseminate the improved cookstoves [18]. It is considered as vital for facilitating a detailed understanding of user needs and paramount for the success of any development program [19]. Involvement of the community takes care of local manufacture, suitability of improved cookstoves to the local cooking practices, and affordable cost [7,20]. However, even these participatory inputs from the user did not always reflect in a favorable outcome for many projects. The term 'participation' includes notions of people's empowerment, involvement, and control over the development initiative [21-23]. Agencies involved in improved

cookstove projects usually have bureaucratic, power-conscious, performance-driven, and goal-oriented structures. It has been very difficult to appreciate people's empowerment and their control in the cookstove development project. Participatory methods adopted by these agencies have been criticized of preserving the participation in terms of mere appearance and actually adhering to a traditional top-down power relation [24]. The Kenyan biomass project in rural areas is an example of such failure of the participatory methods adopted by the Kenyan Ministry of Energy [18] along with donor agencies. Although the improved stove, 'Maendeleo', was very simple and made with locally available technology, not more than 4% of rural households adopted it. Energy saving was not the major concern of the rural dwellers owing to the freely and abundantly available firewood. However, the household income level of the people was very low, and they tended to stick with the traditional practice rather than buying a comparatively expensive (although the stove price was only around US$1.5) alternative. Thus, all the remaining facts can be practically appreciated through people's participation by ensuring their involvement and control in every step.

2. User Priorities of Needs: People tend to adopt interventions on the basis of what matters to them rather than the objectives of the improved cookstove program. The dominance of the interest of outsider stakeholders is one of the reasons of the failures of many improved cookstove programs [18,25]. Improved cookstove projects have been most successful in areas where firewood fetching is an arduous task or people have to spend a significant amount of money on purchasing wood [6]. In both these cases, improved stoves fulfill users' needs. Therefore, the goals of any cookstove program have to be decided on the basis of people's priorities which essentially vary with context.
3. Study of Existing Practices: Cooking practice involves fuel type, fuel size, input power level, number of pots per cookstove, size of pots, time required to cook the food, and ease and flexibility of operation of cookstove. Cooking practices are deeply rooted in the society for hundreds of years [10]. Therefore, an improved cooking method that is different from existing practices has less appeal wherever the people using traditional cookstoves are bound to follow the existing practices because of economic,

social, and religious constraints [26]. Therefore, it is imperative to design a new technology that suits the existing cooking practices and also seeks improvements in the performance. One of the reasons behind multiple failures faced by NPIC and a few market-based companies was that the proposed practice of improved stoves could not match the existing cooking practices carried by the community [10]. Notable examples are dissemination of portable stoves in Maharashtra [27] and Philips/BP-Oorja stoves that accommodate only a specific range of fuel type and size [16]. An improved cookstove should be similar to the traditional one to perform the existing cooking practices [7]. Understanding and accommodation of the users' practice to design an improved cookstove are very important for effective adoption and correct utilization of technology [28], *even at the expense of not achieving the best efficiency*[19,29].

4. Other Uses: Improved cookstoves do not usually cater to a few other uses and benefits that are sometimes derived from cookstoves such as lighting, space heating, reduction of insects and pests, and drying of thatched roof [30]. The design of improved cookstoves should take into account these characteristics of existing cookstoves.
5. Bridging the Gap between Laboratory and Field: One of the most important reasons behind the failure of improved cookstove programs is not meeting the claim to save substantial amounts of fuel. In many cases, improved cookstoves are more efficient than the traditional ones under laboratory conditions, but their performance in field conditions is debatable because improved cookstove designs are found to be incompatible with traditional ways of cooking [30,31]. Oftentimes, the standard Water Boiling Tests (WBTs) are not indicative of the performance of cookstoves in rural communities [32]. This inadequacy can be attributed to lack of stove testing methods which take care of the existing cooking practices [33]. Therefore there, is a need to formulate a performance testing protocol by studying the existing cooking practice to reduce the gap between laboratory- and field-level outcomes.
6. Commercialization and Dissemination: In many successful programs, the stove dissemination has been an independent entrepreneurial activity. Entrepreneurs generally need technical

assistance in designing the stoves and in marketing them to local people [6,29]. A well-designed promotion strategy involves informing the local people about the benefits associated with the improved stoves through a network of local volunteer organizations, television advertisements, or demonstration sites [20]. It is usually found that the local artisans or the trained people seldom adhere to the prescribed dimensions of an improved cookstove [9,16,20]. Thus, for local manufacturing of an improved cookstove, its operation should not be very sensitive to its dimensions.

Overall, it is imperative to understand people's needs and the existing practice to develop an improved cookstove that would be socially and economically acceptable. Improved cookstoves are most popular when they are easily and locally manufactured and have clear advantages in manufacturing cost, fuel economy, matching of existing cookstove practices, ease of use, and durability. Dissemination programs are most effective when they allow for interaction and feedback between stove designers, producers, and users.

Lessons of experiences of early cookstove programs led us to devise a novel approach to address a problem of drudgery involved in firewood fetching activity in a tribal village Gawand wadi (population 293), 120 km from Mumbai, located in Karjat Tribal Block of Raigadh District in the state of Maharashtra in western India (19°04′54.10″ N, 73°27′19.34″ E). This was achieved by interacting with the villagers over a period of 3 years. The study offers three novel contributions. First of all, it proposes an idea of adapted Water Boiling Test for a given context to measure the performance of a cookstove rather than the usual practice of carrying out standard water boiling tests [34-36]. Many researchers achieve an impressive performance enhancement in improved cookstoves after following a standard WBT in laboratory but fail to obtain the same results on the field [30-33]. The present research shows that the results of Kitchen Performance Tests, which follows field-level practice, are close to the results of the modified WBT carried out in the laboratory. The second contribution is a bottom-up approach for cookstove improvement that takes into account the existing cooking practice and brings a progressive increment in the cookstove performance. This approach increases the chances for the adoption of the improvements as compared to many improved cookstove programs (as discussed in the previous subsections) that build a new cookstove (without taking in to account the existing cooking practice) and try to disseminate it in a particular context. It was proven that the combustion and heat transfer of a cookstove can be improved by the introduction of an optimal combination of twisted tapes (with a particular width, number, and twist angle) in the hearth of the cookstove. It effectively enhances heat transfer coefficient by imparting turbulence to the air draft. At the same time, the turbulence also improves combustion by enhancing mixing of air and fuel.

METHOD

The present method adopts a bottom-up approach that emphasizes on understanding the local context and building on it. Such an approach is credited with having a high potential to result in socially and technologically appropriate solutions. Such solutions, in turn, have a greater probability of widespread adoption and long-term sustainability [18]. Towards this end, people's participation, not only in terms of involvement but also in terms of their control in deciding and planning, was sought in every field-level step that was adopted to develop the fuel-efficient cookstove. These steps involved goal setting, study of existing practice, field-level tests, and feedbacks. An emphasis was placed on the study of existing cooking practices after the appreciation of its importance in the present context. To ensure a high degree of social adoption and so to bear minimal variation from the existing cooking practice, it was decided to make small changes in the existing cookstoves by adopting a few guidelines from existing improved cookstove technologies.

Drawing from the abovementioned bottom-up approach, a different methodology was adopted for successful improvement of the traditional cookstove and its commercialization in Gawand wadi. The steps involved in this methodology are described here in short. Each of these steps is elaborated in later sections.

1. Understanding of Present Context: This was accomplished by studying, documenting, and analyzing the village residents from the view point of their assets, activities, energy resources, and utilization of human work hours to carry out different livelihood activities.

2. Goal Setting: The goal of this project was decided by involving the villagers thoroughly. People's participation was ensured with the help of a set of surveys/studies. These included the problem-ranking exercise of the Participatory Rural Appraisal (PRA) activity [37], an energy-timeline survey of the domestic activities (similar to the surveys carried by Date [38] and Reddy[39]), and a health survey.
3. Field-Level Investigation of Existing Practice: This was carried by focused group discussions with women and potters (who make the cookstoves), measurements of existing cookstove dimensions, a walk through the firewood fetching area, and experiments to find the performance of traditional cookstoves. A detailed documentation of the cooking operation was carried out by selecting six households in the village that represent the differences in family size, number of women per unit family, and type of cookstove. The details of the selection procedure are discussed in the section, 'Field-level experimentation'.
4. Adapted Water Boiling Tests: Accommodating the context-specific practices, two types of WBT, namely, single-pot and two-pot WBTs, were designed to measure performance indices of the cookstove.
5. Improvements in the Traditional Cookstove: To ensure minimal variation from the traditional practice (as explained in points 3 and 4 of the Section, 'Learning from the improved cookstove programs'), the traditional cookstove itself was sought to be modified by either optimizing the existing design specifications and/or retrofitting a new device (such as a chimney) in the existing cookstove. The modification was first implemented and tested in the laboratory.
6. Field-Level Experimentation: The proposed modification was tested in the field in the six selected families by adopting the modified water boiling tests and kitchen performance test. To mimic the existing cookstove firing practice, the women from the households themselves fired the cookstoves to carry out field-level experiments.
6. Local Manufacturing and Distribution: Local artisans (blacksmiths) were involved in making any device that needed to be added to the cookstove. An arrangement was also worked out so that the

blacksmith would make and sell these products along with his regular wares (axe, sickle, etc.).

Women's involvement in terms of documentation, suggestions, and feedback at every stage of the process was the crux of the project.

People's Participation

People in the village have contributed in the following ways: (1) people participated in the identification of the projects based upon their need and desires, (2) women participated in the study of the existing cooking practices, (3) women from six selected households (see Table 1) have participated in conducting field-level experiments, (4) women participated in the modification of the standard water boiling test, and (5) women participated in improving the twisted-tape device to make it user friendly. These persons have consented to take part of this study and the publication of the accompanying images.

Table 1: Selected households to carry out field-level experimentation and survey cooking practices

Households	Family size	Number of women	Number of women per unit family size	Type of cookstove
Ambabai Kisan Gawanda	3	2	0.67	Three-pot stove
Tukibai Dhavlya Gawanda	4	1	0.25	Two-pot stove
Gulab Chander Bhagat	5	1	0.2	Three-pot stove
Hirabai Govinda Gawanda	7	4	0.57	Two-pot stove
Shevanta Dehu Bangare	9	4	0.45	Two-pot stove
Bhimabai Kalu Warghade	14	3	0.21	Three-pot stove

Honkalaskar *et al.*

Honkalaskar *et al. Energy, Sustainability and Society* 2013 **3**:16, doi: 10.1186/2192-0567-3-16

Owing to the lack of technical knowhow required to participate in the process of analytical modeling and identification of the retrofitting (twisted-tapes inserts) for the design improvement of the traditional cookstoves followed by its laboratory level design optimization, it was not possible to involve the people in these three activities.

The following section would introduce the context of the study briefly.

Understanding of the Present Context

The major sources of livelihood for the villagers are a single crop rain-fed agriculture (mainly paddy and ragi), collection and selling/consumption of forest produce such as gum, fruits, vegetables, roots, and oil seeds, and agriculture labor in neighboring irrigated area. Occasionally, they also find wage labor work in road construction and infrastructure projects in neighboring areas. The nearest market place is located at a distance of 30 km.

Assets

The average family size is 5.63 with nearly 63% of households between 4 to 8. Percentages of men and women in the population are nearly equal, and the adult population (above 18 years of age) accounts for 71% of the total population. The total village land is 950 ha. Most of the land is sloping and forest land. Total cattle head is 212 with 4.07 cattle head per family, which is higher than the national average of 1.73 [40]. The surrounding forest serves as a source to collect wood fuel, wood, and other forest produce. It comprises both village-owned and government-owned forest land. The main water sources are small earthen dam (around 200 m from the village center, 67,000 m^3 capacity) near the village residential area and three wells.

Activities

The villagers mainly depend on local ecological resources for their livelihood, and therefore, their daily activities vary seasonally. They produce most of the things for their consumption. Their daily activities can be listed as follows:

1. Rain-fed agricultural activities: These involve *Rab* preparation[d], ploughing, puddling, transplanting, weeding, harvesting, threshing, and storage. The Rab activity begins in March, ploughing begins in June (following the first monsoon rain),

other activities follow in the order mentioned, and storage is in October.

[d]This is a pre-monsoon activity in which grass, tree branches, leaves, and cattle dung are burnt over around one-fifth part of the farm to prepare the land for sowing of paddy.

2. Livestock raising: Animal raising is mainly done in the monsoon. In other seasons, animals are set free.
3. Domestic activities: These involve firewood collection, water fetching, cooking, fish/crab catching, plinth preparation, space heating [e], cloth washing, etc.

[e]Space heating is required during winter season to warm the house during night and early morning.

1. Employment and trade: This involves forest collection, wage labor, forest cutting, carpentry, liquor making, sand collection, bamboo work, *moha* seeds collection, brick making, etc.

All these activities involve exchange of time and material and energy resources, and some of them involve exchange of money. The main sources of energy are firewood, animal energy, human energy, cattle dung, grass, kerosene, state-supplied grid electricity, and food.

Time Line and Energy Utilization Survey

Different activities in the village are carried out in a varied manner by each household. An activity-time duration survey showed that the contribution of men, women, and children for all activities (mentioned in the previous section) is 34.00%, 62.88%, and 2.94%, respectively. The percentage distribution of human work hours for different activities is as follows: 29.60% (agriculture), 8.40% (animal raising), 48.50% (domestic), and 13.40% (employment and trade). Domestic activities, which demand most of the human work hours, are mainly carried out by women. Firewood fetching activity demands 14% of the total human work hours spent over all the livelihood activities carried out in the village for 1 year. Using the average human energy demands for different livelihood activities [38], the average per capita energy expenditure in each of these activities is listed in Table 2. Firewood fetching activity requires 9.67% of total human energy expenditure per year.

Table 2: Average human energy expenditure for different livelihood activities per year

Number	Activities	Annual percapita energyexpenditure (Gcal/ year)	Percentage ofenergyexpenditure
1	Animal rearing	15.9	6.43
2	Agriculture	68.0	26.50
3	Water fetching	44.1	17.80
4	Firewood fetching	42.9	9.67
5	Cooking	14.6	8.24
6	Cloth washingand plinthpreparation	9.9	4.03
7	Fishing	3.1	1.25
8	Other activities	64.5	26.04

Honkalaskar *et al.*

Honkalaskar *et al. Energy, Sustainability and Society* **2013** **3**:16, doi: 10.1186/2192-0567-3-16

Goal Setting

The level of people's participation [41, 42] and hence the effectiveness of any technololgy development and dissemination process would increase if the technology intervention addresses a problem that is of high priority to the concerned people. Thus it is imperative to identify the priority that the villagers attach to the need to improve the traditional cookstove.

Perceived Priority of Problems According to Severity

A problem-ranking exercise of PRA was carried out to capture the people's perceptions so that the problems faced by the villagers could be ranked according to their severity. This was achieved by first listing out the problems faced by the villagers followed by individual perceptions about their hierarchy. A total of 37 people including, youth and adult men and women were interviewed. The final hierarchy was found by assigning a weight to each problem equal to the reciprocal of its rank assigned by an individual followed by a weighted average. This study revealed that the firewood fetching activity ranked first among the problems faced by the people. The hierarchy of the problems is shown in Table 3. Presently, both firewood and water fetching activities are carried out by women by carrying loads (around 20 to 32 kg) on their head. Both these activities involve walking on a sloping terrain. These activities have considerable health-related impacts.

Table 3: Overall problem ranking

Problems faced	Young men	Adult men	Adult women	Young women	Overall
Drudgery in firewoodfetching	2	3	1	1	1
Unemployment	1	2	3	7	2
Drudgery in waterfetching	3	5	2	2	3
Lack of healthcare in village	9	1	4	4	4
Low level ofeducation	4	6	8	9	5
Liquor addiction	6	4	7	6	6
Lack of toilets	10	7	5	3	7
Lack of transportationfacilities	5	8	6	5	8
Sporadic electricitysupply	8	10	9	8	9
Snakes bites	7	9	10	10	10

Honkalaskar *et al.*

Honkalaskar *et al. Energy, Sustainability and Society* 2013 **3**:16, doi: 10.1186/2192-0567-3-16

Further, 40 women across various age groups were surveyed to identify the health hazards associated with these two laborious activities. The women could be distributed into the following three approximate age groups: women below 30 years (22), between 30 and 50 years (12), and above 50 years (6). The survey revealed that most the women face problems of backache, neckache, calf muscle ache, and fatigue.

These observations combined with the results of the energy and timeline survey and those from the problem-ranking exercise of PRA were shared with the people in a village-level meeting (December 2009). The villagers came to a consensus that the firewood fetching and water fetching activities were the ones involving the most drudgery and should be tackled first.

The present study focuses on the improvement of the efficiency of the traditional cookstove in order to address the most pressing problem of drudgery associated with the firewood fetching activity without undermining its effect on indoor air pollution. The problem associated with drudgery in water fetching was tackled differently.

Cooking Practice

Design of Traditional Cookstove

There are two types of traditional cookstoves, namely, two-pot and three-pot cookstoves, as shown in Figure 1. There is a common fire input port to feed firewood for both of these stoves. There are 26 two-pot stoves and 28 three-pot stoves in the village. Three-pot stoves are usually found in larger families. Women feel that two-pot stoves are more efficient than three-pot stoves. In the case of three-pot stoves, the third hole is used for a short duration of cooking operation during the day. At all other times, flames come out of this third hole, and heat is lost. Yet, as there is an additional hearth, cooking is faster on the three-pot stoves. The average dimensions of the cookstoves in the village are listed in Table 9 in Appendix 1.

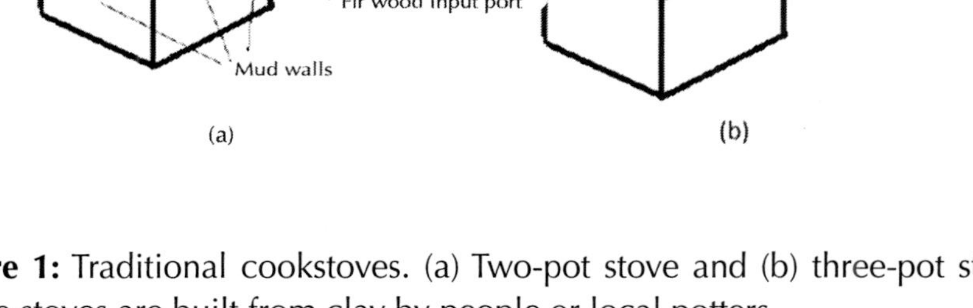

Figure 1: Traditional cookstoves. (a) Two-pot stove and (b) three-pot stove. These stoves are built from clay by people or local potters.

Cookstove Construction

Many households in Gawand wadi make their own cookstove. The remaining households get it manufactured by a few people in the village by offering them a meal. Very few households (presently two) buy cookstoves from potters in nearby villages. The manufacturing process of the cookstoves is elaborated.

Firewood

The socio-technical aspects of firewood (such as firewood fetching and characterization of firewood) are discussed below:

1. Firewood Fetching Activity: Women fetch firewood from the local forest. Every year, this activity starts after paddy harvesting (November) and continues until the end of the next summer. Figure 2 shows an approximate area from which women fetch firewood in Gawand wadi. The dotted line indicates a path travelled by women during winter season to fetch firewood. It is typically 4 to 5 km. The thick line indicates the firewood fetching path during summer season. It is typically 6 to 7 km. The time required to fetch a firewood bundle varies from 3 to 4 h in winter to 5 to 6 h in summer. Initially, tree branches, fallen during monsoon storms, are collected. The area traced per day to collect these branches increases gradually, thereby increasing the daily time required to fetch a firewood bundle. Eventually, when it is very difficult to get fallen wood branches, small-diameter (<5 cm) tree branches are cut. This branch cutting operation demands more human energy than just collection. The typical weight of a firewood bundle carried by women varies from 20 to 32 kg depending upon the load carrying capacity of the woman and the firewood requirement of the household. Figure 3 shows the undulated elevation profile of a path travelled by women to carry firewood during winter. Firewood fetching activity can be divided in to five subtasks. These are to reach the forest area, to collect firewood, to tie the bundle of wood with a thin stem of a creeper or bark of a few tree species, to cut tree branches if required, and to carry firewood home. A typical time breakup required to carry these tasks is shown in Table 4.

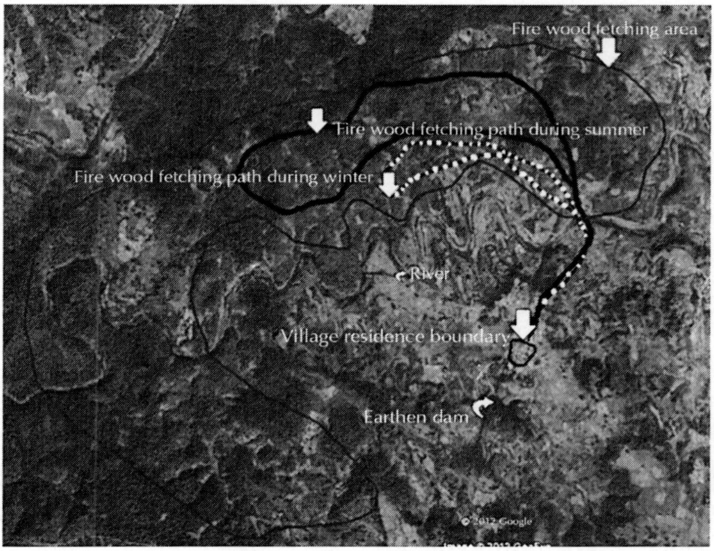

Figure 2: Firewood collection area in Gawand wadi [43]. An approximate area from which women fetch firewood in Gawand wadi. It spans around 10 km^2. The dotted line indicates a path travelled by women during the winter season to fetch firewood. It is typically 4 to 5 km. The thick line indicates the firewood fetching path during the summer season. It is typically 6 to 7 km.

1. Wood species which are used as firewood are *ain* (*Terminalia elliptica*), *dhamada* (*Angeissus latifolia*), *chera* (*Erinocarpus nimmonii*), *karvanda* (*Carissa carandas*), *bondari* (*Lagerstroemia parviflora*), *nana* (*Lagerstroemia microcarpa*), *akashi* (*Salix babylonica*), *kuda* (*Holarrhena pubescens*),*sag* (*Tectona grandis*), etc. Out of these, *dhamada, ain,* and *nana* give a good flame. An approximate usage pattern of different firewood species perceived by villagers is shown in Table5.
2. Properties of Firewood: Firewood species which are mainly used for cooking were characterized by finding seasonal moisture content, calorific value, and proximate analysis as listed in Table 5. The calorific value was measured using a bomb calorimeter. The proximate analysis was carried using an electric furnace. A dry wood sample (around 1 g) was heated in an electric furnace at 550°C for 7 min and then at 750°C for an additional 2 min to find the percentage of volatile matter in the wood species by measuring the percentage reduction in the weight. It was further

heated at 750°C for 4 h to find the percentage of carbon in the sample by calculating the reduction in weight. Tabulated values of calorific values and proximate analysis are averages of three such experiments with less than 5% of variation. Moisture content was measured by the following method: A fixed quantity (100 ± 20 g) of wood was dried in an electric oven at 105°C for 24 h. The fraction of moisture was calculated by subtracting the fraction of dried wood mass. The moisture content varies with season. During rainy season, the moisture content for 3-cm diameter *ain* wood is around 18% to 20%. It varies from 8% to 14% during summer and winter seasons.

3. Typical Firewood Dimensions: The firewood diameter lies in the range of 1 to 5 cm. To obtain a good flame, firewood of 1.5 to 4 cm diameter is used. For the initial ignition of fire, wood sticks below 1.5-cm diameter are used. Firewood of diameter larger than 5 cm is usually chopped. The length of firewood lies in the range of 50 to 70 cm.
4. Wood Requirement Per Year: This varies with family size. For an average family size of six, the wood requirement per year is around 2,700 kg.

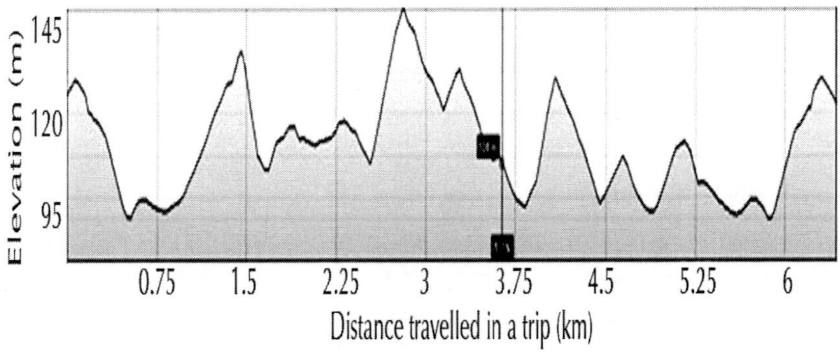

Figure 3: Firewood fetching path during the winter season in Gawand wadi [43]. Elevation of a firewood fetching path traced during winter as shown by dotted line in Figure 5. It shows that the path is quite undulated. The difference between the highest and the lowest location is around 50 m.

Table 4: Approximate time breakup for different tasks performed in firewood fetching activity

Tasks	Average timerequired inwinter (h)	Average timerequired insummer (h)
To reach forest area	0.5	0.7
To collect wood sticks	2	4
To cut tree branches	0	3 (wood eitheris collectedor cut)
To tie bundle with a climber	0.25	0.25
To carry firewood bundle home	0.75	1
Total	3.5	5 to 6

Honkalaskar *et al.*

Honkalaskar *et al Energy, Sustainability and Society* 2013 3:16, doi: 10.1186/2192-0567-3-16

Table 5: Wood species wise usage percentage and calorific value

Wood species	**Perceived usage in percentage**	**Calorific value (MJ/Kg)**	**Percentage of volatiles**	**Percentage of carbon**
Ain (*Terminalia elliptica*)	50	16.8	76.5	22.3
Dhamda (*Angeissus latifolia*)	20	17.9	72	26.5
Chera (*Erinocarpus nimmonii*)	10	17.9	-	-
Kuda (*Holarrhena pubescens*)	10	17.3	-	-
Sag (Tectona grandis), karvanda (Carissa carandas), bondari (Lagerstroemia parviflora), akashi (Salix babylonica), and others	10	*sag* **(16.6),** *karvanda*(16.6), *bondari* **(17.1)**,*akashi* **(17.3)**	-	-

Honkalaskar *et al.*

Honkalaskar *et al. Energy, Sustainability and Society* 2013 **3**:16, doi: 10.1186/2192-0567-3-16

Cookstove-Firing Practice

This was studied by observation of actual cooking, a questionnaire survey, measurement of coal formation and temperature at different parts of the hearth zone, and direct experience of cookstove firing. The following are the observations:

1. Kerosene is used for the initial ignition of firewood. Comparatively dried and smaller diameter wood is preferred for the initial burning. It usually takes 2 to 3 min to achieve a steady-state flame.
2. Usually, two or three larger diameter wooden sticks (4 to 6 cm) and a few smaller diameter sticks (2 to 4 cm) are fired simultaneously to attain the required heating rate.
3. In general, 50 to 70 cm long wood sticks are preferred so that the fire can be adjusted easily by slight movement of these sticks. Firewood is arranged in such a manner that sufficient primary air can pass through the fuel feed port.
4. If necessary, the women blow air by mouth to maintain a good flame size.
5. Red hot charcoal helps sustain the flame. It is necessary to remove the excess charcoal if it is accumulated in the hearth zone to make space for the firewood.
6. The woman who is cooking sits on one side of the cookstove to avoid exposure to radiated heat from the wood insertion hole.

Durability

On an average, a cookstove lasts for 2 to 3 years before its disintegration. Usually, its fuel feeding hole disintegrates first. It wears due to frequent friction between firewood sticks and its edges. Pot rests also disintegrate as these wear due to regular loading and unloading of pots

Food Preparations

There are three types of food preparations, namely, baking (*bhakari* or roasted bread), boiling (heating water, rice, curry, and tea), and frying (vegetables dishes and fish/crab curry). Table 6lists the average food preparation time, the pot size, and the duration of operation for a two-pot cookstove for a family of six. It shows that nearly 32% of the operations are carried in the single-pot mode, and 68% in the two-pot mode for a two-pot cookstove. Correspondingly, for three-pot stoves, the percentage of operation in the two-pot mode and the three-pot mode is 54 and 14, respectively.

Table 6: Average daily cookstove practice of an average family of size six

Cookstovep-reparations	Per capitaper mealreq-uirement	Durati-on(min)	Number ofdailypre-parations	Totalprepa-rationtime (min)	Potdiam-eter(cm)	Potheight (cm)	Single-potmode (min)	Two-potmode (min)	Woodcons-umption(kg)
Waterboilingfor bathing	4.2 l	21 ±2	5	105	27.5 ± 1.5	16 ±2	42	63	2.9
Tea	76.3 g	10 ±1	3	30	14 ±1	7 ±1	10	20	0.75
Rice	141 g	24 ±4	2	50	19 ±2	12 ±2		48	1.3
Bhakri	162 g	24 ±10	Once in3 days	8	26 ±2	5		8	0.2
Dal**a**	24 g	35 ±5	2	80	17 ±2	9 ±2		70	1.7
Bhaji		12 ±3	2	24	26 ±2	5		24	0.62
Total							55	115	7.4

[a]Here, dal is referred to as *tur dal* (yellow pigeon pea) and *kala ghevda* (a Maharashtrian bean).

Honkalaskar *et al.*

Honkalaskar *et al.Energy, Sustainability and Society* 2013 3:16, doi:10.1186/2192-0567-3-16

Input Power of Cookstove

The input power is the energy input to the cookstove per unit time. It varies with recipes, cookstove dimensions, family size, and cookstove firing practice. The capacity of a cookstove increases with the size of cookstove or the volume of hearth zone. Usually, tea brewing and rice making requires lesser amount of input power. Other practices such as water boiling require higher amount of power input to the cookstove. If the number of women in the family is low (one or two), they tend to reduce the time required for cooking by firing the cookstove at higher input power in order to spare more time for other activities.

Power input to the cookstove is evaluated by two different ways as follows:

1. The ratio of hearth zone volume to input power of cookstove follows a rough rule of thumb of 0.6 l/kW [44]. Using an average hearth zone volume of 9.4 l (for two-pot cookstove), the power input to the cookstove is 15.6 kW if it is used in the two-pot mode. For single-pot mode operation, the average hearth zone volume is 6.5 l, and thus, the estimated power input is 10.8 kW.
2. Water boiling tests carried out on the selected cookstoves yielded an average input power of around 14.6 kW for two-pot mode of operation and 8.46 kW for single-pot mode of operation. This value is closer to the input power (14 kW) measured by Geller [45] in the village of Ungra near Bangalore.

Experimental Procedure

Table 6 shows that around 40% of the daily firewood is utilized in water boiling for bathing. It is the operation consuming the most amount of firewood. Other activities including tea brewing and rice and curry making are close to the water boiling operation, and these activities cumulatively comprise more than 85% of the cooking operations. Therefore, a WBT and a kitchen performance test (KPT) [36,46,47] that involves a real time evaluation of the daily cooking practice were used to the water boiling test, a fixed amount of water is boiled on a cookstove, and the efficiency is calculated by dividing the total heat transferred to water by the total heat input.

The components of the cookstove system are operator, firewood, cookstove design, pot, and type and amount of food to be cooked. Each component plays a role to influence the performance of a cookstove. Therefore, the water boiling test was modified to accommodate existing cooking practice defined by these components. The modifications are listed below:

1. To mimic existing cookstove firing practice, it was decided to let women fire the cookstove while carrying WBTs on field (see Figure 4).
2. Time duration of WBT: Most of the cooking preparations in the village last for a little less than half an hour as documented in Figure 5. Therefore, it was decided to carry WBT for half an hour.
3. Fuel size: For field-level testing, wood of 2 to 5 cm diameter and 50 to 70 cm length was used, the wood species being mainly *ain* and *dhamda*. For laboratory level testing, wood of 2.5 ± 0.5 cm diameter and 25 ± 2 cm length was used.
4. Power input: 8.5 kW (1.81 kg firewood/h) in single-pot mode and 14 kW (3.2 kg firewood/h) in two-pot mode of operation.
5. Weight of water: 5.5 l (average water usage for the cooking preparations).
6. Pot size: 26 to 31 cm (range of pot sizes used in the village to boil the bathing water).
7. Timing to carry out tests: 11 am to 5 pm.
8. Starting condition: Hot start, owing to the hot start condition for most of the cooking preparations. This step was added after obtaining feedback from the women.

Figure 4: A woman firing cookstove in a two-pot water boiling test.Water boiling tests in the village were carried out with women, where a woman who cooks food fired her cookstove to find out its performance.

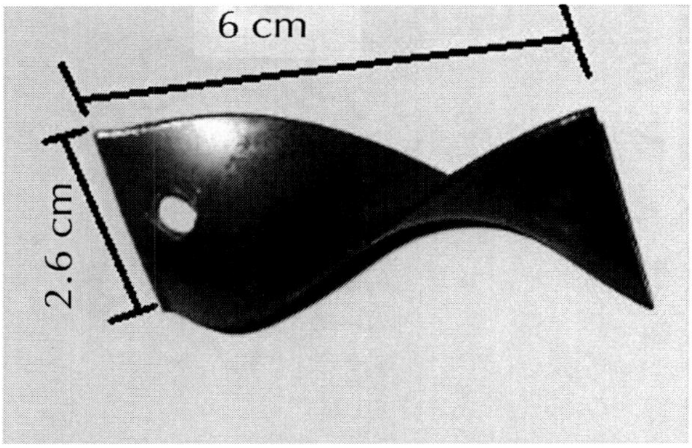

Figure 5: A twisted mild steel plate.

Kitchen Performance Test

KPT measures the average rate of firewood consumption by a stove in a normal household environment [36]. A wood pile (mainly including the wood species *ain* and *dhamda*) of 100 ± 15 kg was stacked in a household, and the rate of consumption of firewood per household was found by dividing the total consumed weight of wood by the number of days of its consumption.

Proposed Improvement

Cookstove operation comprises three phenomena that occur simultaneously, namely, (a) burning of fuel, (b) mass transfer of air, volatile gases, and gaseous products of combustion, and (c) heat transfer to pot and other parts of the stove [48, 49]. The overall thermal efficiency ($_t$) of the cookstove depends upon three intermediate efficiencies, namely, combustion efficiency of fuel ($_c$), heat transfer efficiency to pot ($_h$), and efficiency at which pot transfers energy to food ($_p$). The efficiency value $_p$ depends upon pot size which is usually determined by food type. The thermal efficiency can be improved by either optimizing the design specifications or retrofitting a certain device (such as a chimney) in the existing cookstove or both to improve heat transfer efficiency and/or combustion efficiency.

To study the cookstove operation for its design improvement, a steady state analytical model was developed by splitting the hearth into three zones to study char combustion, volatile combustion, and heat transfer to the pot bottom separately. A study of the variation of design specifications on the thermal performance revealed that the traditional cookstove specifications are close to their optimal values. A Sankey diagram of the cookstove operation revealed that the major thrust areas to improve the thermal performance are the heat transfer to the pot and combustion of volatiles[f].

[f]The details of the thermo-chemical model of the traditional cookstove are not discussed in the present report and will be presented elsewhere.

It has been documented that the incorporation of twisted tapes (TTs, see Figure 5) in the hearth generate swirl motion of the gases which improves the jet impingement heat transfer [50,51] at the pot bottom. It also improves combustion of volatiles by increasing air-fuel mixing and the residence time of the reactant gases [52,53]. Bhandari et al. [47] found that the inclusion of twisted tapes improved the efficiency of an experimental cookstove. It was decided that the effect of the inclusion of twisted tapes to improve the thermal performance of the traditional cookstove be investigated.

RESULTS AND DISCUSSION

Effect of Variation of Number of Twisted Tapes on Thermal Performance of Cookstove

Manufacturing in the Laboratory

The twisted tapes were manufactured by twisting a mild steel plate heated to 300°C. The strip dimensions used for laboratory tests were the following: for the width, 1.25, 2.54, and 3.38 cm; length, 6 cm; and thickness, 1.5 mm. The tapes were twisted by holding them in a fixture together with a lathe machine. To ensure smooth twist, the tapes were heated with a torch blower, and the twisting procedure was implemented in four to five steps.

Retrofitting of Twisted Tapes

TTs were retrofitted in a cookstove by just inserting them in two or more steel rods and keeping the rods on the holes of the hearth to let the TTs hang in the hearth as shown in Figure 6. Thus, the intervention does not change the existing cookstove design. The new cookstove is similar to the existing one and would address the concerns raised in point numbers 3 and 4 of the Section "Learning from the improved cookstove programs".

Figure 6: Twisted tapes hanging in the hearth holes. This shows the way to hang the twisted tapes in the hearth holes by using steel rods. They are placed

in a way that they cover the cross-section of the hearth evenly.

Outcomes of Laboratory Tests

A two-pot cookstove, made by a local potter, was used for the laboratory-level experimentation. Its specifications are listed in Table 10 in Appendix 1. The two-pot water boiling tests were carried out by varying the number of TTs in each hearth (experimental set-up is shown in Figure 7). The water boiling tests were carried out in the following order:

1. Varying the number and width of the twisted tapes in each hearth for the same twist angle (180°).
2. Varying the twist angle (0°, 60°, 120°, 180°, and 240°) for optimum width and number of twisted tapes (2.54 cm and seven, respectively).
3. Carrying a few additional WBTs with optimum width and twist angle (2.54 cm and 180°, respectively) to study the effect on exhaust gas composition and soot. This helps in understanding the contributions due to the twisted tape assembly.

Figure 7: Experimental set-up to carry two-pot WBT in the laboratory.This shows the experimental set-up that was used to carry out the laboratory WBTs. Four thermocouples were used to measure the temperature of water and flame for both the pots and hearths.

It was observed that the thermal efficiency of the cookstove was optimum (improvement by 24.5%) for inclusion of seven number of twisted tapes having a width of 2.54 cm and a twist of 180°. A few additional WBTs with optimum width and twist angle showed that the proposed retrofitting improves combustion of the volatile gases (see Table 7). It was also observed that the amount of soot accumulated over the outer surface of the pot decreases by around 38% (see Table 7) by inclusion of these TT inserts. A detailed analysis of the laboratory level WBTs is not discussed in the present report and will be presented elsewhere. It was decided that field-level tests by retrofitting seven TTs of 2.54 cm width and 180° twist angle in each hearth be carried out.

Table 7: Soot accumulation and percentage composition of the gases at the center of the stove top

	O_2(%)	$\frac{CO}{CO}$	Soot (g)
Normal cookstove	10.5	0.0872	2.84
Seven TTs	5.5	0.082	1.75
Eleven TTs	7.8	0.080	2.20

The listed values are averages of three WBTs.

Honkalaskar *et al.*

Honkalaskar *et al.* *Energy, Sustainability and Society* 2013 **3**:16, doi:10.1186/2192-0567-3-16

Field Level Experimentation

Field-level experiments were carried out at the six selected households in the following order: single-pot water boiling test, two-pot water boiling test, and kitchen performance test. The method of selecting the households is explained below.

Selection of the Households

It was first noted that the cookstove size and firewood requirement increase with family size. If the number of women per unit family size is higher, firewood fetching and cooking activities get distributed among women, and it reflects in the human work hour involvement and drudgery associated with it. If the number of women per unit family size is lower, women try to fire the cookstove with a higher input power in order to accomplish the cooking task earlier. Figure 8 shows all the existing combinations of the family size and number of women per unit family size in the village. The data points are then divided in four quadrants as shown. Depending on the willingness of the family members to participate in field-level experimentation and the type of cookstove (two-pot or three-pot), four different households from each quadrant were selected along with two households from extreme ends (encircled in Figure 9 and listed in Table 1). As there are nearly equal number of two-pot and three-pot stoves in the village, it is ensured that three out of the six selected households have three-pot stoves. The dimensions of the cookstoves are listed.

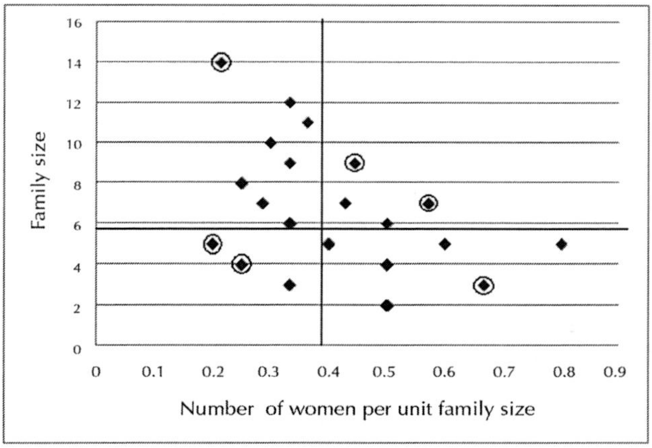

Figure 8: Family size versus number of women per unit family wise in Gawand wadi. The graph shows all possible combinations of family size and women per unit family size. The sample space is divided in four quadrants for even selection of the households across such variations. Encircled families were selected.

Figure 9: TT pack. This shows a twisted tape pack that can be retrofitted in the traditional cookstove.

Single-Pot Water Boiling Test

By introducing seven TTs, the thermal efficiency of the cookstove increases by an average of 20.61 ± 1.98%, and the specific fuel consumption of cookstove is decreased by 14.5 ± 1.32% (at the statistical significance level of 0.05 for both the outcomes).

Two-Pot Water Boiling Test

The thermal efficiency of the cookstove increases by an average of 22.82 ± 2.48%, and specific fuel consumption of the cookstove decreases by 15 ± 1.29% (at the statistical significance level of 0.05 for both the outcomes) by retrofitting of TTs in existing cookstove (see Figure 10 and 11).

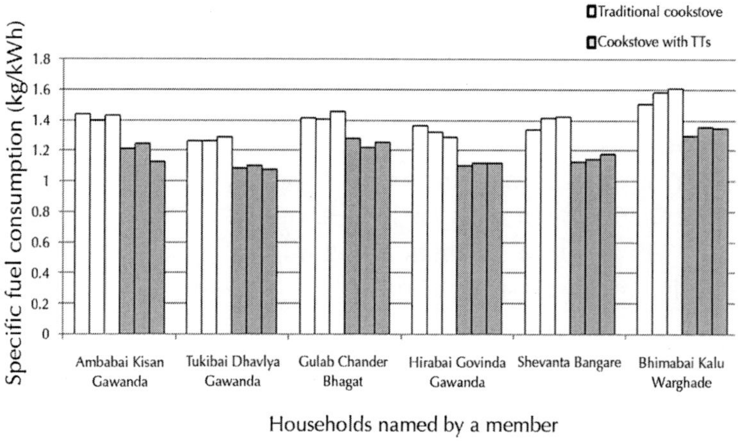

Figure 10: Effect of retrofitting of TTs on thermal efficiency of traditional cookstoves in Gawand wadi. This shows the results of the two-pot water boiling tests that were carried out in six selected households. Traditional cookstove and cookstove with inclusion of TTs were tested by carrying out three tests for each kind of cookstove. It can be observed that the thermal efficiency of the cookstove increases by the inclusion of TTs for every household.

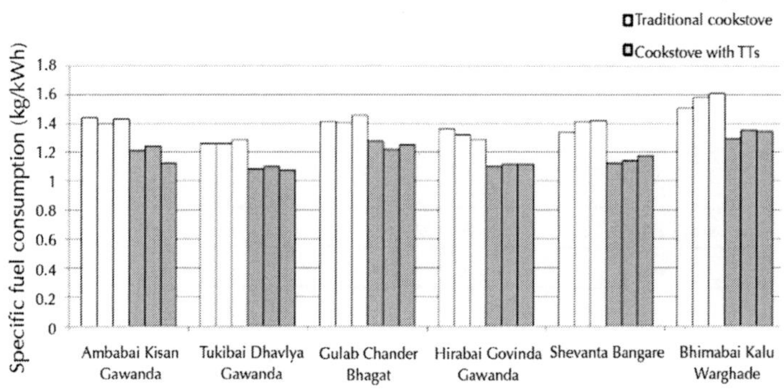

Figure 11: Effect of retrofitting of TTs on specific fuel consumption of traditional cookstoves in Gawand wadi. Figure shows results of the two-pot water boiling tests that were carried out in six selected households. Traditional cookstove and cookstove with the inclusion of TTs were tested by carrying out three tests for each kind of cookstove. It can be observed that the specific fuel consumption of the traditional cookstove decreases by the inclusion of TTs for every household.

Kitchen Performance Test

The results of the KPT are shown in Table 8. Corrections in wood consumption of the families due to presence of guests were incorporated by considering an average firewood usage for food preparations as listed in Table 6. The results of KPT show (see Table 8) a reduction of 21.3 ± 1.89% (at the statistical significance level of 0.05) in the average daily wood consumption.

Table 8: Results of kitchen performance test at two households

Household	**Operation conditions**	**Average daily wood requirement (kg/day)**
Ambabai Kisan Gawanda	Normal cookstove	5.81
	Cookstove with TTs	4.56
Tukibai Dhavlya Gawanda	Normal cookstove	5.78
	Cookstove with TTs	4.39
Gulab Chander Bhagat	Normal cookstove	7.25
	Cookstove with TTs	5.56
Hirabai Govinda Gawanda	Normal cookstove	7.22
	Cookstove with TTs	5.89
Shevanta Dehu Bangare	Normal cookstove	9.08
	Cookstove with TTs	7.43
Bhimabai Kalu Warghade	Normal cookstove	9.39
	Cookstove with TTs	7.38

Honkalaskar *et al.*

Honkalaskar *et al. Energy, Sustainability and Society* 2013 **3**:16, doi:10.1186/2192-0567-3-16

The results of the kitchen performance test show that the percentage increase in the performance of cookstove is lower than that indicated by WBTs. Once the cooking preparations in the morning, noon, or evening are completed, the women leave accumulated coal inside the hearth itself. It slowly burns and decreases in weight until the next operation starts. Secondly, if the coal accumulation prohibits insertion of firewood in the hearth zone, the women take the coal out of the cookstove and throw it away. Thus, all the coal that remains at the end of cooking operations is not utilized. Hence, the percentage decrease in wood consumption is lower (by around 3%) than the percentage increase in the thermal efficiency. Yet, as some amount of the remaining coal is still utilized for further food preparation, the percentage decrease in average wood consumption is higher in KPT than the percentage decrease in specific fuel consumption (by around 4%) in WBT.

As KPT gives actual improvements in the cookstove performance by enabling real-time monitoring of the cooking performance, it is a more relevant test. Furthermore, as water boiling is a major cookstove operation (see Table 6), the results shown by the water boiling tests are closer to those shown by the kitchen performance test.

Women's Feedback

The women in the six selected households gave the following feedbacks regarding the retrofitting of TTs in their cookstoves:

1. Improvement in the cookstove has reduced wood consumption by roughly 25% (as per their estimate).
2. Soot accumulation over the outer surface of pots has reduced.
3. Time required to prepare different food items has reduced.
4. Women found that sometimes the steel rod holding the TT would fall inside the hearth zone due to lateral force exerted by firewood on TTs. If this problem is removed, it would be wonderful.

A new retrofitting was devised as a result, in which the seven TTs are inserted in a welded triangular rod structure as shown in Figure 9. It does not fall in the hearth and is very easy to place and remove from the stove. Women liked this new design.

Commercialization through Local Artisans

Commercialization involves manufacturing and distribution of the technology. Bottom-up approach emphasizes building on the local human skills and the local socio-economic structure. The improvement is found to be not very sensitive to the slight changes in the width and twist angle of the twisted tape. Thus, local manufacturing and dissemination of the TT packs was found to be an impressive alternative. The success of the dissemination of an improved cookstove program depends upon multiple factors [54] that vary with context. Although many dissemination programs emphasized on training local people for manufacturing and marketing of improved cookstoves[20,55], it was found that at many places, these enterprises get poor benefits, so they have little motivation to start and sustain the business [9]. Therefore, if a local artisan is trained to accommodate the manufacturing and selling of improved cookstoves in his/her existing socio-commercial business, there is a greater probability for its sustenance. In the present context, manufacture and dissemination of the TT packs through the existing business structure of the local blacksmiths is sought. Traditionally, they make and repair sickles, axes, ploughs, and other tools required for agricultural and domestic activities. Recently, many of them have diversified their skills to lay tin roofs and build steel gates. Manufacturing of TT pack involves cutting, drilling, heating, twisting, and welding operations. Thus, the local blacksmiths who have welding and drilling machines can manufacture the TT pack. There is a traditional custom of buying and repairing the iron/steel tools from local blacksmiths. Thus, the manufacture and selling of the TT packs can be accommodated in the existing sustainable traditional business of the blacksmiths.

A blacksmith's work load is seasonal. He makes and repairs ploughs and sickles during monsoon for agricultural activities, while he makes axes and other tools in the winter season. During summer, he is engaged in house building activities that involve welding, tin/steel cutting, and drilling operations. Usually, a blacksmith has leisure time during monsoon season. Thus, the blacksmiths can allot more time to make TT packs during monsoon.

A manufacturing process to build the TT packs has been devised by interacting with two blacksmiths and one steel fabricator [g].

[g]Madhukar Chavan from Kashele, Anil Jadhav from Kondivade, and Pankaj Gore from Vaingaon.

1. A steel plate with thickness of 1.5 to 2 mm is cut to get small plates of required TT size.
2. The plates are placed in the blacksmith's kiln to make them red hot.
3. The red hot plate is held in a vise and twisted by a long lever with a groove to hold the plate. It is twisted by 60° to 90° in one go.

It is then again placed in the kiln to heat. The whole process of twisting it to 180° requires two to three steps. This is to ensure a smooth curve for the twisted tape.

4. A hole is drilled in the TT to insert a steel rod (4 mm diameter).
5. Finally, a TT pack is made by welding the rods together.

The present material cost is US$0.5 per TT pack. By considering the manufacturing cost, transport cost to buy steel plate and rods, and the profit of blacksmith (US$8 per day), the total cost comes to around US$1.25 per TT pack. The technology is being disseminated with the help of two local volunteer organizations, Disha Kendra and Oak Vanaushadhi Kendra.

Dissemination of the Innovation

As of now, 45 families from 11 different villages [h] have been using the TT packs in their cookstoves. Ten out of these families have been using this device for more than a year. Similar traditional cookstoves are found across the coastal belt of Maharashtra. Thus, the same innovation can be disseminated across the villages in this coastal region. For the other parts of rural India, the stove designs are much different, and there is a need to identify a context-specific intervention, which may not necessarily be a twisted tape.

[h]These villages are Gawand wadi, Tepachi wadi, Waghai wadi, Kothimbe, Rajpe, Kalamb, Vare, Kondivade, Tamhanath wadi, Khandpa, and Dhare wadi located in Karjat block of Raigadh district of Maharashtra state.

CONCLUSIONS

The present study derives a novel bottom-up approach to develop or improve a rural/community technology by appreciation of wider aspects of the technology-community linkage, namely, people's priorities, people's desires behind improvement or development of technology, importance of studying and maintaining existing technology practises, availability of local resources to manufacture and operate the technology, and its commercial viability and sustainability. It enhances social efficiency of adoption of the technology by people. Overall, the approach mingles three aspects linked with rural/community technology.

1. Social: This includes people's participation to set the goal behind the development of technology, appreciation of the existing context in terms of livelihood activities, local natural resources, and human skills.
2. Technical: This includes the formulation of a procedure to measure the performance of the technology according to existing context, laboratory and field-level testing, feedback from the people, and further improvements in the technology.
3. Commercial: Manufacturing and distribution through locally sustainable structure.

Retrofitting of the TT pack in the existing cookstoves in the village would save nearly 22% of human work hours and human energy involved in the firewood fetching activity. For an average family of six, women make 97 to 122 trips per year to the local forest to fetch firewood. These trips can be decreased by 21 to 27. By considering an average duration of 4.5 h/trip, there can be an annual saving of 95 to 122 women work hours. The TT pack also decreases soot accumulation on the outer surface of pot by around 38%. Based on the temperature variation graph of the water boiling test in the laboratory, the TT pack reduces the cooking time by around 18.5%. For an average family of six, it saves 0.5 h daily and 182 h annually required for cooking preparations. Although the laboratory experiments showed that the optimal retrofitting of the twisted tapes improves combustion and thereby reduces indoor air pollution, it remains to measure the percentage reduction of health-damaging emissions [i].

[i]Major health-damaging emissions involve particulate matter, carbon monoxide, carcinogens such as benzo[a]pyrene, formaldehyde, nitrogen dioxide, and sulfur dioxide [59,60].

AUTHORS' CONTRIBUTIONS

VHH contributed in capacity development, carrying out field-level investigations, field tests, and report writing. He is thus the lead contributor to the paper. UVB and MS have contributed in providing guidelines to devise the methodology and providing inputs on the experiments and in the overall sense to structure the report. All authors read and approved the final manuscript.

ACKNOWLEDGEMENTS

The authors are thankful to the people of the village Gawand wadi, especially to the women, for their participation and support throughout the research work.

REFERENCES

1. Brown E (1996) Deconstructing development: alternative perspectives on the history of an idea. J Hist Geogr 22(3):333-339
2. Smith KR (1989) Dialectics of improved stoves. Econ Polit Weekly 24(10):517-522
3. Westhoff B (1995) One of the oldest technologies in the world - from the open hearth to the microwave. In: Westhoff B, Germann D (eds) Stove images: a documentation of improved and traditional stoves in Africa, Asia and Latin America 1995, Frankfurt: Brandes and Apsel.
4. Sesan TA (2011) What's cooking? Participatory and market approaches to stove development in Nigeria and Kenya. University of Nottingham: Doctoral thesis.
5. Crewe E (1997) The silent traditions of developing cooks. In: Grillo R, Stirrat R (eds) Discourses of development: anthropological perspectives, Oxford: Berg.
6. Barnes DF, Openshaw K, Smith KR (1994) What makes people cook with improved biomass stoves? A comparative international review of stove programs. Washington DC: The World Bank: Technical report, World Bank Technical Paper Number 242, Energy Series.
7. Sanchez T (2008) Lessons from project implementation on cook stoves and rural electrification, the practical action experience. Rugby: Technical report, Practical Action.

8. Smith KR, Shuhua G, Kun H, Daxiong Q (1993) One hundred million improved cookstoves in China: How was it done? World Dev 21(6):941-961
9. Jagdish KS (2004) The development and dissemination of efficient domestic cook stoves and other devices in Karnataka. Curr Sci 87(7):926-931
10. Krishna C (2000) Improved cook stoves; yet to be a success story. http://bioenergylists.org/files/IMPROVED_COOK_Stoves_Evaluation Accessed 21 Feb 2012
11. Venkataraman C, Sagar A, Habib G, C NL, Smith K (2010) The Indian National Initiative for Advanced Biomass Cookstoves: the benefits of clean combustion. Energ Sus Dev 14:63-72
12. Kishore V, Ramana P (2002) Improved cookstoves in rural India: how improved are they? A critique of the perceived benefits from the National Programme on Improved Chulhas (NPIC). Energ 27:47-63
13. Bailis R, Cowan A, Berrueta V, Masera O (2009) Arresting the killer in the kitchen: the promises and pitfalls of commercializing improved cookstoves. World Dev 37(10):1694-1705
14. Lim SS, Vos T, Flaxman AD, Danaei G (2013) A comparative risk assessment of burden of disease and injury attributable to 67 risk factors and risk factor clusters in 21 regions, 1990 2010: a systematic analysis for the global burden of disease study 2010. Lancet 380(9859):2224-2260
15. Rai K, McDonald J (2009) Cookstoves and markets: experiences, successes and opportunities. London: Technical report, GVEP International.
16. Dhoble R, Bairiganjan S (2009) Cooking practices and cook stoves field insights: a pilot study of user experience with traditional and improved cook stoves. Technical report, Centre for Development Finance, Institute for Financial Management and Research. India: Chennai.
17. Smith KR (2010) What's cooking? A brief update. Energ Sus Dev 14:251-252
18. Sesan TA (2012) Whose priorities? Evaluating objectives of participatory development in the context of household energy

projects in Africa. http://www.gbengasesan.com/temidocs/WhosePriorities.pdf. Accessed 23 Oct 2012

19. Rouse J (1999) Improved biomass cookstove programs: fundamental criteria for success. The University of Sussex: Masters thesis, The Centre for the Comparative Study of Culture, Development & the Environment.
20. Ergeneman A (2003) Dissemination of improved coosktoves in rural areas of developing world: recommendations for the Eritrea dissemination of improved stoves program. Eritrea: Technical report, Eritrea Energy Research and Training Center.
21. Oakley P (1998) Strengthening people's participation in rural development. New Delhi, India: Society for Participatory Research in Asia, Occasional Paper Series No. 1
22. Pearse A, Stiefel M (1980) Inquiry into participation: a research approach. Geneva: Technical report, United Nations Research Institute for Social Development.
23. Westergaard K (1986) People's participation, local government and rural development: the case of West Bengal, India. Copenhagen: CDR research report 8, Centre for Development Research.
24. Parfitt T (2004) the ambiguity of participation: a qualified defence of participatory development. Third World Q 25(3):537-555
25. Troncoso K, Castillo A, Masera O, Merino L (2007) Social perceptions about a technological innovation for fuelwood cooking: case study in Rural Mexico. Energ Policy 35:2799-2811
26. Shrestha SK, Thapa R, Bajracharya K (2003) National improved cook stove dissemination in the mid-hills of Nepal, experiences, opportunities and lesson learnt. Yogyakarta, Indonesia: Technical report, The Asia Regional Cookstove Program Network.
27. Sinha B (2002) The Indian stove programme: an insiders' view - the role of society, politics, economics and education. Boiling Point 48:23-26
28. Ruiz-Mercado I, Masera O, Zamora H, Smith KR (2011) Adoption and sustained use of improved cookstoves. Energ Policy 39:7557-7566

29. Rouse J (2002) Community participation in household energy programmes: a case-study from India. Energ Sus Dev 6(2):28-36
30. Quadir SA, Mathur S, Kandpal TC (1995) Barriers to dissemination of renewable energy technologies for cooking. Energ Policy 36(12):1129-1132
31. Tata Energy Research Institute (1989) Evaluation of improved stoves in Tamil Nadu, Rajasthan and West Bengal. New Delhi: Technical report, Submitted to the Department of Non-conventional Energy Sources, New Delhi, Tata Energy Research Institute.
32. Berruetaa VM, Edwardsb RD, Maserac OR (2008) Energy performance of wood-burning cookstoves in Michoacan, Mexico. Renew Energ 33:859-870
33. Makonese T, Chikowore G, Annegarn HJ (2011) Potential and prospects of improved cookstoves (ICS) in Zimbabwe. Cape Town, 11–13 April 2011: Paper presented at domestic use of energy (DUE) conference.
34. Bailis R, Ogle D, MacCarty N, Still D (2007) The water boiling test. Technical report, Shell Foundation.http://ehs.sph.berkeley.edu/hem/content/WBT_Version_3.0_Jan2007a.pdf
35. Bailis R, Ogle D, MacCarty N, Still D (2004) The Water Boiling Test (WBT), version 1.5. Technical report, University of California-Berkeley.
36. Volunteers in Technical Assistance InternationalStandards (1985) Testing the efficiency of wood burning cook stoves. Virginia, USA: Technical report, VITA.
37. Chambers R (1994) The origins and practice of participatory rural appraisal. World Dev 22(7):953-969
38. Date AW (1989) Energy utilization pattern of Shilarwadi. Indian J Rural Technol 1:33-63
39. Reddy AKN (1982) Rural energy consumption patterns - a field study. Biomass 2:255-280
40. National Sample Survey Organization (2010) Press note on income, expenditure and productive assets of farmer households (January–December 2003). http://pib.nic.in/release/rel_print_page1.asp?relid=14531 *webcite*. Accessed 2 Jan 2012

41. Arnstein SR (1969) A ladder of citizen participation. J Am I Plann 35(4):216-224
42. Aref F, Redzuan M (2009) Assessing the level of community Participation as a component of community capacity building for tourism development. Eur J Soc Sci 8:68-75
43. Google (2012) Maps. http://maps.google.com/ *webcite*. Accessed 20 March 2012
44. Lepeleire GD, Krishna Prasad K, Verhaart P, Visser P (1981) A woodstove compendium. Netherlands: Technical report, Wood burning stove group, Eindhoven University of Technology.
45. Geller HS (1982) Cooking in the Ungra area: fuel efficiency, energy losses, and opportunities for reducing firewood consumption. Biomass 2(6):83-101
46. Bhatt MS (1982) the efficiencies of firewood devices (open fire stoves, chulhas, and heaters). Indian Acad Sci 5(part 4):327-342
47. Bhandari S, Gopi S, Date AW (1988) Investigation of CTARA wood-burning stove. Part 1. Experimental investigation. Sadhana 13(4):271-293
48. Baldwin SF (1952) Biomass stoves: engineering design, development, and dissemination. 1600 Wilson Boulevard, Suite 500, Arlington, Virgnia, 22209, USA: Technical report, VITA.
49. Food and Agriculture Organization of The United Nations (1993) Improved solid biomass burning cookstoves: a development manual. Bangkok: Technical report, Regional Wood Energy Development Programme In Asia, GCP/RAS/154/NET, Field Document No.44.
50. Nuntadusit C, Waehahyee M, Bunyajitradulya A, Eiamsa-ard S (2012) Heat transfer enhancement by multiple swirling impinging jets with twisted-tape. Int Commun Heat Mass 39:102-107
51. Kinsella C, Donnelly B, O'Donovan TS, Murray DB (2008) Heat transfer enhancement from a horizontal surface by impinging swirl jets. Paper presented at the 5th European thermal-sciences conference, The Netherlands, 18–22 May 2008.
52. Guillaume DW, LaRue JC (1995) Combustion enhancement using induced swirl. Exp Fluids 20:59-60

53. Benajes J, Molina S, Garcya JM, Riesco JM (2004) The effect of swirl on combustion and exhaust emissions in heavy-duty diesel engines. J. Automobile Eng 218:1141-1148
54. Ramirez S, Dwivedi P, Bailis R, Ghilardi A (2012) Perceptions of stakeholders about nontraditional cookstoves in Honduras. Environ Res Lett 7(4):2-11
55. Karve P (2005) A model for dissemination of improved biomass fuels and cooking devices through rural enterprises. Boiling Point 50:26-28
56. The World Bank (2011) Household cookstoves, environment, health, and climate change: a new look at an old problem. Washington, DC: Technical report, The World Bank.
57. A Volunteer Group of Foresters in Nepal (2013) Improved cooking stoves. http://www.forestrynepal.org/wiki/117 *webcite*. Accessed 8 May 2013
58. Wikipedia (2013) Cook stove. http://en.wikipedia.org/wiki/Cook_stove. Accessed 8 May 2013
59. Bruce N, Perez-Padilla R, Albalak R (2002) the health effects of indoor air pollution exposure in developing countries. Switzerland: Technical report, World Health Organization Geneva.
60. World Health Organization (2010) WHO guidelines for indoor air quality: selected pollutants. Copenhagen, Denmark: Technical report, World Health Organization, Regional Office for Europe.

Citations

CHAPTER 1

Shu Kee Lam, Deli Chen, Arvin R. Mosier, and Richard Roush, the Potential for Carbon Sequestration in Australian Agricultural Soils is technically and Economically Limited, doi: 10.1038/srep02179.

CHAPTER 2

Gamfeldt, L. et al. Higher levels of multiple ecosystem services are found in forests with more tree species. Nat. Commun. 4:1340m, doi: 10.1038/ncomms2328 (2013).

CHAPTER 3

Vandana Sharma and Smita Chaudhry, "An Overview of Indian Forestry Sector with REDD+ Approach,"ISRN Forestry, vol. 2013, Article ID 298735, 10 pages, 2013 doi:10.1155/2013/298735.

CHAPTER 4

Devan Allen McGranahan, Aaron L. Daigh, Jessica J. Veenstra, David M. Engle, James R. Miller, and Diane M. Debinski, "Connecting Soil Organic Carbon and Root Biomass with Land-Use and Vegetation in Temperate Grassland," The Scientific World Journal, vol. 2014, Article ID 487563, 9 pages, 2014, doi:10.1155/2014/487563.

CHAPTER 5

Maria Nijnik, Arie Oskam, and Anatoliy Nijnik, "Afforestation for the Provision of Multiple Ecosystem Services: A Ukrainian Case Study," International Journal of Forestry Research, vol. 2012, Article ID 295414, 12 pages, 2012. doi:10.1155/2012/295414

CHAPTER 6

XiuHai Zhao, Javier Corral-Rivas, ChunYu Zhang, Hailemariam Temesgen, and Klaus v Gadow, Forest observational studies-an essential infrastructure for sustainable use of natural resources, doi:10.1186/2197-5620-1-8.

CHAPTER 7

Michael P Perring, Rachel J Standish, and Richard J Hobbs, Incorporating Novelty and Novel Ecosystems into Restoration Planning and Practice in the 21st Century, doi:10.1186/2192-1709-2-18.

CHAPTER 8

Vijay H Honkalaskar, Upendra V Bhandarkar, and Milind Sohoni, Development of a Fuel Efficient Cookstove through a Participatory Bottom-Up Approach, doi:10.1186/2192-0567-3-16.

Index